BRAINSTORM

ALSO BY SUZANNE O'SULLIVAN, MD

Is It All in Your Head?

SUZANNE O'SULLIVAN, MD

Brainstorm

DETECTIVE STORIES FROM THE
WORLD OF NEUROLOGY

Other Press
New York

First published in the United Kingdom in 2018
by Chatto & Windus, an imprint of Vintage.
Vintage is part of the Penguin Random House
group of companies.

Production editor: Yvonne E. Cárdenas

Typeset in India by Integra Software Services Pvt. Ltd, Pondicherry

10 9 8 7 6 5 4 3 2 1

LIBRARY OF CONGRESS CATALOGING-IN-PUBLICATION DATA

Names: O'Sullivan, Suzanne.
Title: Brainstorm : detective stories from the world of neurology /
Suzanne O'Sullivan, MD.
Description: New York : Other Press, 2018. | Includes index.
Identifiers: LCCN 2018015388 (print) | LCCN 2018023195 (ebook) |
ISBN 9781590518670 (ebook) | ISBN 9781590518663 (hardback)
Subjects: LCSH: Neurology—Popular works. | Neurology—Case
studies. | BISAC: HEALTH & FITNESS / Diseases / Nervous System (incl.
Brain). | SCIENCE / Life Sciences / Neuroscience. | PSYCHOLOGY /
Neuropsychology.
Classification: LCC RC346 (ebook) | LCC RC346 .O88 201 (print) |
DDC 616.8—dc23
LC record available at https://lccn.loc.gov/2018015388

For Aisling Kellam
and E.H.

The possession of knowledge does not kill the sense of wonder and mystery. There is always more mystery.

—Anaïs Nin

CONTENTS

THE BRAIN

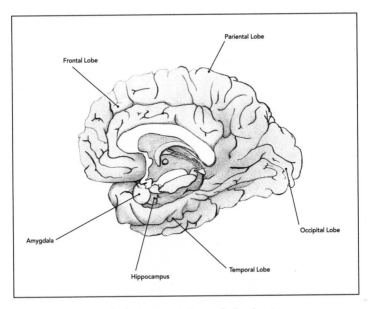

Medial cross-section of the brain

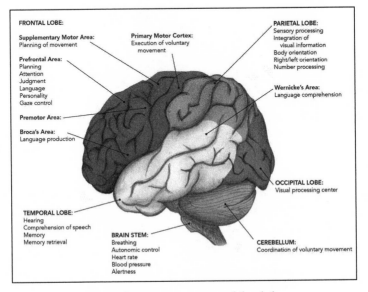

Brain function segregated by lobe

INTRODUCTION

The brain is a world consisting of a number of unexplored
continents and great stretches of unknown territory.

—Santiago Ramón y Cajal (1852–1934),
pathologist and neuroanatomist

There were three doctors working at the clinic with over fifty
patients to be seen. I was the most junior doctor. John, the
senior registrar, was in a room next to mine. He was a few years
ahead of me in training. Experience counts for a great deal as
a doctor so his knowledge was far greater than mine. The third
doctor was the consultant for whom we both worked.

As usual there were far too many people to be seen in the
time allotted. We all had to work more quickly than we were
comfortable with. I was required to discuss any difficult problem
with John or the consultant. It was a time in my career when
I thought that a good doctor was one who worked quickly and
wasn't a nuisance to their senior colleagues. I avoided asking
for help if at all possible.

The patients' notes were piled high on a trolley outside the
consultant's room. They were visible to all the anxiously waiting
people. Everybody turned to look as I took a set off the top

of the pile and brought them into my office to read. They contained only a few pages. I was relieved. A thick file meant years of history to understand and a chronic problem that might not be solvable. So many neurological conditions are incurable and very challenging to treat. A thin file could indicate a minor issue that had disappeared since the patient was last seen. When I opened the notes, though, I sighed. The man had only been to the clinic once before and I was the doctor who had seen him that time too. The tests I had ordered then came back as normal, which meant I hadn't found the source of the problem. I would have preferred one of the other doctors to see him this time. Maybe they would notice what I had missed.

My letter recording our conversation said that he had odd sensations in his right arm. I had examined him and found nothing amiss. I had wondered if the problem might be a trapped nerve in his neck. I ordered electrical tests to examine the integrity of the nerves traveling into the arm. Those tests found that the nerves seemed to be working as they should. I knew that if the man was no better since that first meeting I wouldn't really know what to do next. My only hope was that he had recovered without my help. I called him into the room.

"How have you been?" I asked.

"The same," he told me, and my heart sank.

"Okay. Well ... can you explain the problem to me again?"

"I get goosebumps running down my right forearm. Very noticeable goosebumps. That's all of it."

He made it sound so simple but the symptoms just didn't speak to me.

"Is there any numbness?" I asked.

"No."

"In between the goosebumps, does your arm feel otherwise normal?"

"Mostly, except when the goosebumps arrive."

He opened and closed his fist and stared at the offending arm. I was trying to feel my way through the problem. Trying to understand. I was not succeeding.

"Is your hand or arm weak?"

"No … maybe … no. When I have the goosebumps it feels weird enough that I think, if I was holding something, I would drop it."

"How often do you get the goosebumps?"

"Only for a minute or two once a day. Maybe twice."

The man was in his thirties. He looked well and had no medical problems in his past. I wondered why he was so worried about symptoms that only lasted one minute a day. What he described seemed almost trivial to me.

"Well, the good news is that the tests are totally normal," I told him. "I don't think you have anything to worry about."

I was revving up my reassuring speech, hoping that he was one of the worried well. Maybe all he needed was to be told that everything was okay.

"But what is it then?"

Oh no. His voice was anxious. A normal test result was not good news to him. He wanted a better explanation than I could give.

"I'm not convinced that what you describe can be fully explained … but most symptoms that can't be explained just disappear when left alone. I mean ... goosebumps? Is it the temperature in your office ... the air conditioning?"

I was clutching at straws and we both knew it.

"I don't think you understand," his voice was getting higher, "they are goosebumps that stand up like anthills on my skin. It isn't normal ... it's ... it's ... unnatural."

I blush easily when I am uncomfortable and out of my depth. I felt the redness crawl up my neck towards my face. I felt goosebumps creeping over my own skin.

"Let me look at your arms again," I suggested, buying myself some thinking time. I asked him to sit on the couch and take off his shirt. I looked at his muscles and they seemed normal. I tapped his reflexes with my tendon hammer and they were normal too. I poked at his arm with a blunt pin to check sensation. Normal. I tested his strength. It was possible his grip was not as strong in the right hand as the left but I had a sense that he wasn't trying very hard. Perhaps he needed me to find something wrong.

"I don't think I can explain this," I said eventually.

Just for a second I thought I saw him roll his eyes. I took that as my cue that I needed help.

"If you don't mind waiting I need to go and discuss the problem with my consultant," I told him.

"Thank you," he said, obviously relieved.

As I walked across the corridor towards my boss's office I cringed. I didn't want to interrupt him just to ask his advice about a man with occasional goosebumps.

I knocked lightly on the door and it swung open.

"Aah, here's trouble," John said as he beckoned me in, laughing. He had also come to the consultant to discuss a patient.

Our relationship was one in which he teased me about every small thing I missed, and reminded me of them as often as possible. I made sure to get my own back at every opportunity. We liked each other. Rivalry is a part of working in medicine. Mistakes, even understandable ones, tend to be remembered.

I closed the door behind me.

"Can I get your advice on this man?" I asked, indicating the notes in my hand.

"How many more patients are out there?" the consultant asked.

We had all been holed up in our rooms working as quickly as possible, but with no sense of how many people we had seen between us.

"There's quite a big pile of notes left on the trolley," I told him, "but can you please talk to this guy I'm seeing? I'm not sure what to do with him. He has goosebumps in his right arm, but that's it. I thought maybe a radiculopathy. I sent him for nerve-conduction studies last time but the results were normal. Should I scan his neck, maybe? I'm not sure because it's not clearly dermatomal. There's nothing to find on exam."

Dermatomes are one of the many anatomical patterns of the nervous system that neurologists use to trace a patient's symptoms to a location in the network of nerves and spinal cord and brain. They refer to the areas of the skin known to receive their nerve supply from the spinal nerve roots. The skin of the arm is divided into seven dermatomes. If you have altered sensation in a single dermatome – a patch over your shoulder or hand, for example – then it implies a lesion in a specific spinal nerve root. I had not been able to make proper sense of my patient's problem. The odd feelings in his arm did not fit into one neat dermatomal package, but it was as close as I could get. I had focused my investigations on the spinal nerve roots wondering if he had a trapped nerve in his neck. The tests told me I was wrong.

"Did you talk to me about this man the last time he was here?" the consultant asked.

"Yes."

At every clinic I saw as many patients as I could. I discussed the difficult cases with the consultant as soon as I had seen them so he had the chance to see them too. The people who were more straightforward we discussed when the clinic was over. Of course, this system meant that the consultant depended entirely on my judgment and on the quality of the information I gave him.

The consultant, John and I walked back to my clinic room. I thought I heard the other patients in the waiting room heave a collective sigh. They were still watching the pile of notes, waiting for their turn. When none of us picked up a new set they knew they were in for a further delay.

The consultant introduced himself to my patient. "So I've been hearing about these odd feelings in your arm. Can you explain it to me again, if you don't mind?"

The man didn't mind. He looked relieved to be seen by a more serious-looking doctor.

"So I get this slow wave of goosebumps pass over my skin and then it's gone."

He ran his hand over his forearm to indicate which area was affected.

"How long does it take for the wave to pass?"

"I guess about a minute. Maybe less. It is an awful feeling. Really horrible."

"Does it feel the same every time?"

"Yes."

"What does the arm feel like at other times?"

"Not quite right. I can't fully explain it."

"And everything else is okay? The other arm? The legs? No headaches or anything else I should know about?"

"Nothing."

The consultant took his ophthalmoscope and moved close to the patient to look at the back of his eyes. Then he tested the strength and sensation in the limbs.

"Right hand maybe a bit less strong than the left?" He looked over his shoulder at me as he spoke.

"I wasn't sure," I answered.

"Does it happen day and night?" the consultant asked the patient.

"It can happen anywhere. I can wake in the middle of the night with it or I can get it walking down the street. Always exactly the same. Do you know what's causing it?"

"Not immediately but I think we need to organize some further tests. This doctor will arrange for you to have a scan of your brain and we'll see if that provides some answers," the consultant nodded his head in my direction. Turning back to the patient he offered some reassuring words and promised we would be in touch very soon. As he left the room he said quietly so only I could hear, "Well young lady, it seems you were looking in the wrong place!"

A week later the magnetic resonance imaging (MRI) brain scan result was available. The temporal lobes are part of the brain, running front to back along the side of the head at the level of the ear. In this man's left temporal lobe nestled a brain tumor. The tumor was too small to cause headaches. All it was doing was irritating the surrounding cortex, the gray matter that makes up the surface area of the brain. The cortex is electrically active. Through disrupting it the tumor was causing unwanted bursts of electrical activity. The result of these autonomous brainstorms were epileptic seizures. The only manifestation of these seizures was piloerection – goosebumps.

I had missed a brain tumor. I had done so by making two mistakes. Firstly, I hadn't listened properly. Patients usually offer the diagnosis to their doctor in their story. Diagnosis relies on the doctor's ability to appreciate the subtleties of what they are being told. When my patient described the odd feeling in his arm I thought he was telling me about a sensory disturbance – a problem primarily involving the nerve pathways that carry touch to the brain. But goosebumps are not strictly a sensory disturbance and are not carried by the sensory nervous system. They are an autonomic phenomenon. Part of our flight and fight response, they are one of the manifestations of fear and arousal. The autonomic nervous system is an entirely different confluence of nerves than those that detect pain or touch. Being a neurologist is being a detective. To find the cause of a neurological problem you must start by figuring out the pattern and then search in the right anatomical region. You interpret clues and follow them. By misunderstanding the clues I had followed them to the wrong place.

My second mistake was to underestimate the breadth of brain disease. I had not scanned the brain because I forgot what a devious organ it can be and how heterogeneous the manifestations when it is diseased. There is a tendency to associate disorders of the brain only with the most obvious symptoms – paralysis, memory loss, headaches, dizziness, blackouts. But the brain plays a role in the function of every single organ in the body, every muscle movement (voluntary or involuntary), every tiny gland, every hair follicle. When things begin to go wrong in the brain it stands to reason that anything in the body can go wrong as a result. Not just the big things, the small things too. Brain disease may cause a flagrant symptom like paralysis or it may pick off one minuscule function. In my patient's case the brain lesion was so small that it had stimulated the center for autonomic control and nothing else. Thus goosebumps were the only symptom of a brain tumor.

For a doctor it is always terrible to get a diagnosis wrong. I take some consolation in reminding myself that when I was first studying medicine in the 1980s this man's tumor would have been too small to be detectable by the technology we had then. Nor did goosebumps appear in the index of any neurology book I owned. For a very long time the practice of clinical neurology was limited by how difficult it was to investigate the brain. Diagnosis was educated speculation, with no way to gather the evidence that proved the neurologist's detective work right or wrong. More than most people realize, despite major technological advances, this is still the case. The brain, the seat of what makes us human, is still vastly uncharted territory. And neurology remains one of the most complicated and beguiling of all medical disciplines.

*

Historically the brain has presented a bigger challenge to science than any other organ. The heart beats. The lungs inflate and deflate. But the brain comes with no superficial clues as to how it works. Encased in bone, it is uniquely hard to get at. Even when that barrier has been passed there is nothing to indicate which part of the brain does what. Engaged in the most complex activity, it remains inert to the naked eye.

Detailed pictures of the gross anatomy of the brain had been available since the eighteenth century. All based on postmortem examinations, of course. Anatomists had divided the brain into brainstem, cerebellum and cerebrum. The cerebrum comprised four lobes: frontal, temporal, parietal and occipital. Scientists could only guess at the specific function of these structures. Or even if they had a specific function.

Noting that the hand and foot had each been carefully designed for their purpose scientists surmised that the brain was also shaped to fit. Inspection of the cerebrum revealed that the hills and valleys (gyri and sulci) of its surface had a very similar pattern in every person. Because the cerebrum was "soft" and the cerebellum "hard" they speculated that the former was responsible for sensation and the latter motor function; and that discrete brain areas might have a predetermined purpose. This was all guesswork. There was no way to verify it except to wait for people to suffer injury or illness and observe the results.

Many of the early vital discoveries in neurology are attributable to a single individual, sometimes a doctor and sometimes a patient. Of all the patients, Phineas Gage is the most famous. He took us on our earliest baby steps into the brain.

In 1848 Gage, a railway worker, was injured in an explosion. A tamping iron was blown through his skull destroying his left frontal lobe. The accident turned Gage from being a quiet man to one prone to aggression. It was the first clue to the role that the frontal lobes play in our lives. It was through Gage's inadvertent frontal lobotomy that science first suspected the importance of the frontal lobe to personality.

For a very long time war wounds, suicide attempts, accidents and strokes were the only investigative tools of the neurologist. Doctors hovered over the injured and dying to learn. At first this was a very random method for accruing knowledge, but in time it became more organized with the development of the clinical–anatomical method, a systematic way to delineate the typical features of neurological diseases. Neurologists examined patients in life, followed them until death and then correlated the clinical picture with what was found at autopsy. By comparing lots of patients doctors learned to distinguish the clinical features of a spinal cord disorder from that of the brain. Or the pattern of limb weakness that occurs in a nerve disease from that of a muscle disease. Key clinical signposts emerged. Scraping the bottom of the foot and producing an upward movement of the big toe meant a brain or spinal cord disorder. Tapping the reflex points and finding the reflexes absent indicated a possible problem in the peripheral nerves.

The clinical–anatomical method was the beginning of the practice of neurology as we know it now. Through it neurologists learned how to recognize patterns of disease based on the careful search for clinical signs. It taught us how to correlate particular disabilities with anatomical locations in the brain.

But a system that relied on accidents and postmortems would never provide all the answers. A window was needed into the living brain. At the end of the nineteenth century that window appeared in an unexpected form. It was not an innovation or anything new, it was something ancient – the disease epilepsy.

Epilepsy was noted as a brain disease by Hippocrates in 400 BC. It took millennia for this to be fully accepted and longer still for the mechanism through which seizures develop to be understood. But once it was understood the unique brain lessons epilepsy could offer were quickly harnessed.

The story of how epilepsy became the most vital investigative tool for the brain began with the Italian scientist nicknamed "the frog's dancing master." In the eighteenth century Luigi Galvani demonstrated that the biological cell has electrical properties. He electrically stimulated the legs of frogs and made their muscles twitch. This was the start of research into the electrical signals and patterns emitted by nerves, muscles and the brain.

A hundred years later neurologist John Hughlings Jackson was watching an experiment in which a colleague was stimulating the cortex (the outer layer of the brain) of a monkey. Hughlings Jackson was a doctor who subscribed to the method of learning through close observation. Something in the animal experiment struck a cord of recognition in him. It reminded him of an epileptic seizure. In his assessment of patients with seizures he had noticed that muscle jerking often spread systematically through the body, starting in one place and working its way to another. He saw a similar march of symptoms when he watched the animal experiments. Hughlings Jackson decided epilepsy was caused by a sudden disorderly expenditure of force spreading through the brain. He later expanded that view to say that the spreading force was an electrical discharge. He believed, correctly, that the discharge began in the cortex and spread through the connections between cells. The symptoms produced by the progression of that discharge evolved according to the function of the cells affected. This theory fitted well with the suspicion that different parts of the brain represent different parts of the body and that the arrangement of the brain is very similar in each of us.

Suddenly an epileptic seizure had become a symptom rather than a disease. The features of each seizure were representative of the part of the brain engulfed in the offending electrical discharge. If the discharge affected the area of the brain thought to control the right side of the face, then it caused twitching there. If the discharge then spread to the region controlling the right hand it followed that the twitching spread to involve that hand. Thus, watching a patient have a seizure was like taking an anatomical tour through the brain.

This theory led to neurologists and neurosurgeons joining forces to correlate brain lesions with seizure symptoms. For example, if a patient had seizures that caused their arm to jerk and an operation demonstrated a tumor in his frontal lobe it seemed reasonable to surmise that the part of the brain where the tumor lay must have some importance to the motor control of the arm. Doctors were following seizures to their source and extrapolating the function of the brain from that. Through comparing patients it was possible to make rudimentary maps putting symptoms together with brain regions.

Of course this technique had similar limitations to the clinical–anatomical method. It depended on chance occurrence. Also if the surgeon opened the skull and didn't immediately see the source of the problem they had no way of knowing where else in the brain to look for it. A way to search the normal brain was needed. Epilepsy also provided that. Seizures have a unique feature that make them particularly useful as a tool to probe the brain. They can be reproduced artificially using the technique of neurostimulation.

Since the end of the nineteenth century scientists had been electrically stimulating animal brains without apparent harm to the animal. The advent of anaesthetics and antibiotics meant that this technique could also be applied to humans. The brain itself has no sensory receptors and can be touched, cut into and stimulated without causing pain. With the use of local anaesthetics surgeons opened the skulls of conscious patients. They then electrically stimulated the cortex. Being fully conscious the patients could report what each cortical stimulation made them feel. In animal studies researchers could only observe the reaction in the animal, but a person could describe their experience. Some brain stimulations produced movement. Others caused sensory disturbances, hallucinations, the resurgence of memories or emotional upset.

Most of the patients who were operated on had epilepsy. By testing different areas of the brain the surgeon searched for the source of their seizures by trying to reproduce their symptoms. For example, if a person's seizure began with the experience of a hallucinatory smell the surgeon stimulated different cortical regions until the patient reported the experience of that smell.

If they succeeded it was assumed that the source of the seizure had been found. It was also assumed that that area must have some importance to the normal processing of smell. But scientists did not have to restrict the use of this technique to looking for pathology. They began to use it to methodically investigate the function of healthy brain tissue. By systematically stimulating different areas of cortex and recording the results, surgeons were able to get a much better understanding of how the cortex was arranged. New discoveries no longer needed to rely entirely upon random injury and disease. Neurostimulation allowed them to create more sophisticated functional maps of the brain.

Fast forward to my early training years. A century had passed and the vast majority of neurological diagnoses were still made on an entirely clinical basis. A major development had come in the 1970s, in the form of the CAT scan, which allowed us to look inside living organs for the first time. It brought with it an opportunity to confirm a clinical diagnosis at an earlier stage in some patients. It could detect tumors and strokes invisible to us before. But it had limitations; there were still many pathologies it could not see. The mysteries of the brain were still not even close to being unraveled. The neurologist's powers of deduction, their ability to interpret their patients' stories, were all. They drew on their knowledge of neuroanatomy and brain maps to make a diagnosis, and medical investigations were mostly supplementary.

I was already a junior doctor, training in neurology, when MRI scans were introduced to most hospitals in the mid-1990s. MRI showed the brain in astonishing detail and, unlike CAT scans, did not give a dose of radiation with every scan. This meant that they were safe to be used regularly on the same person. Even a child's delicate, developing brain could be imaged without fear of consequences. MRI could be used to find pathology, but also to track the changes of the normal growing brain.

While CAT scans and MRI scans were significant medical advances, it is important to realize that both were only still pictures – photographs. They showed structural anatomy, but said nothing at all about brain function. Staring at an MRI scan told you no more about how the brain works than staring at computer circuitry tells you how a computer processes information. Awake, asleep, engaged in a complex mental activity, MRI pictures of the brain all look the same. It was not until the twenty-first century that new imaging techniques enabled the investigation of brain *function* as well as structure. But there is still no technology that can predict or explain intelligence, talent, compassion or humor. No scan can tell a doctor how their patient feels. The peripheral nervous system can be anatomically unpicked, identifying which nerve goes to which muscle or organ, but the tightly packed brain is not so easy to deconstruct. Technology is helpful but the clinical assessment still trumps any test results.

I qualified as a doctor in 1991 and as a specialist in neurology in 2004. The years I spent in training turned out to be a very exciting time in the field of neuroscience. As well as imaging technology becoming more precise and leading to a better understanding of how the different regions of the brain work together, there were also many pivotal discoveries in the field of genetics.

These gave new insights into neurological disease and the normal workings of the nervous system. They also offered the opportunity to make some diagnoses with the use of a single blood test. But these advances have not improved the lot of those affected by brain disease as much as people might think. The development of new treatments has not kept pace with our advances in knowledge. We still don't know what causes the majority of brain diseases, nor have we figured out how to reverse them. There are still far more unknowns than knowns where the brain is concerned. How is personality determined? How is information processed? It is very hard to interpret and mend brain disease when we are still trying to understand its basic biology.

*

I do not ever remember doubting that neurology was the specialty for me. The nervous system is beautiful. It is intricate. Tiny nerves run up and down the limbs and through the spinal cord, all ultimately communicating with the brain through billions of long threadlike axons. Nerves coalesce in some places and separate in others, all by careful design. Each carries its own specific message and travels its own predetermined route. The complexity of it all allows us to function in a highly sophisticated way. When things go wrong that same complexity means neurological diseases often feel as if they have an almost infinite way of manifesting their symptoms. A centimeter to the right or left, the same tiny tumor in the spinal cord or brainstem or brain will produce an entirely different picture.

Medical students often find neurology very intimidating. Walk into an exam and encounter a patient who is losing weight, has wasted muscles in their hand and a drooping eyelid and the average medical student will baulk. To a junior neurology doctor, already becoming versed in how the puzzle pieces of the nervous system are arranged, the exact same problem is an easy one. They know that in the region of the shoulder, at the apex of the lung, there is a bundle of nerves. Included amongst them are some that go into the arm and one that ultimately travels to the eye. Cancer growing in that uppermost part of the lung can invade the nerve bundle, causing weakness in the hand and eyelid. The challenge of tracking these sorts of signs and symptoms is exactly what draws many neurologists to the specialty. As a medical student I found this process as daunting as anyone else, but I was also sufficiently intrigued to want to know how it was done.

I now work as a consultant neurologist in the field of epilepsy. The twenty-first century has brought many new tools that allow me to interrogate the function of my patients' brains – but the art of neurology remains the same. I still do exactly what the pioneers did – extrapolate from a patient's account of their symptoms to a diagnosis. I draw from their description of their experience to a location in the brain. I interpret stories. The maps of my predecessors and modern technology have made that process much more accurate, but many patients' problems still defy the knowledge available. We are always learning. The symptoms of brain disease are open to such endless possibilities that the search for answers is nowhere near over. The scope and effects of brain disease are as great as the scope of the healthy brain.

Neurological diagnosis is solving a puzzle, but when you rarely have all the pieces. You are given ten pieces of a hundred-piece jigsaw and are asked to predict the final picture from just those. Even today, nobody knows what a completed map of the brain will look like, so many puzzles are impossible to fully solve.

Goosebumps are far from the most challenging clinical case I have faced. They were just the beginning. In this book I will share the strange stories of some of the other people who have tested my knowledge, often to its limits: a stressed school janitor who hallucinated a fairy-tale scene; a ballet dancer who couldn't stop falling; an office worker who lost her trust in the person she loved; a girl who kept running away. Joan of Arc and Alice in Wonderland will feature, alongside some very brave people facing radical brain surgery to cure a disability that you probably wouldn't even notice if you met them. I will show how medical advances coexist with, and still entirely depend upon, old-fashioned medicine. All of the people you will hear about have seizures, but no two will be the same. Epilepsy has provided some of history's greatest insights into the brain. These people will show you how and why.

This is a book about the brain, epilepsy and humanity. It is about the incredible strength and ingenuity of people who have disabilities that are uniquely their own. Doctors have always learned from patients. It is my belief that the patients in *Brainstorm* have much to teach us all.

1

WAHID

I am about to discuss the disease called "sacred." It is not, in my opinion, any more divine or more sacred than other diseases, but has a natural cause, and its supposed divine origin is due to men's inexperience and to their wonder at its peculiar character.

—*On the Sacred Disease*, attributed to Hippocrates, *c.*400 BC

I went to the waiting room and called Wahid's name. In one corner there was a bustle of activity. Lids being put on coffee cups. Coats and bags being gathered. I held the door open and waited. A couple started towards me. They didn't make it very far before the man turned back towards his chair and ran to claim some forgotten gloves. The woman waited for him. In the background the receptionist smiled over at me.

"I caught you by surprise by calling you on time!" I said when the couple finally made it into the room. My attempt at humor didn't dent their troubled expressions.

"Is it okay if I come in with him? I'm his wife," the woman said.

"Of course," I said. "I'm Dr. O'Sullivan, by the way."

There was a further brief dance with bags and chairs before we all sat down. Wahid's slim notes sat on the table between us. They contained only one letter. All it said was, *Please see this man who has been woken by strange attacks at night since he was twelve years old.* I looked at the man sitting across from me. Outwardly he was the picture of health. Young, tall, well built, neatly dressed. I looked at his date of birth. Twenty-five. Whatever was bothering him had been doing so for a very long time but hadn't outwardly touched him.

"So, Wahid, your doctor tells me that some sort of odd attacks are waking you at night," I said opening the notes on a blank page, ready to record everything that was said. "But perhaps before we get to that you could confirm for me how old you are and if you are right- or left-handed."

Every question matters. Patients are strangers. At the start I am as interested in how they answer as I am in the answer itself.

"He's twenty-five and he's right-handed," Wahid's wife said.

"Are you working? Studying?" I asked.

There was a brief whispered exchange in a language I couldn't place.

"Shh," his wife said and then turned back to me. "He's in college."

"What do you study?"

Another rumble of half-whispers prompted me to interrupt, "Do you speak English?"

The referral letter did not say otherwise, but it seemed necessary to check.

"He speaks perfect English ... he just doesn't want to be here," she said and raised a quietening hand to her husband who looked ready to voice some objection.

"I see you've had this problem since you were a child. What prompted you to come here now after all this time?"

I had been wondering this since I read the referral letter. I directed the question pointedly at Wahid, trying to force him to answer for himself.

"Me," his wife said wearily, "I made him come."

"If it helps at all, I won't make you do anything you don't want to," I assured him. In neurology more than any other specialty there are so few absolutes. Most consultations are a collaboration. Some are a negotiation. "Let's go through the story and see what's needed – if anything. So ... this thing ..." I hesitated to know how to refer to these mysterious as yet unchronicled events. "This thing that happens at night – do you remember the very first time it happened?"

When a patient goes to see their doctor they often describe their symptoms as they are on the day of the visit. Or the day before. Or the worst day. Their pain at its peak. A doctor needs to know these details, but like any other story the ending may be misleading if you don't know what led the person there. The first symptom is the first puzzle piece.

"Actually I do remember how it started."

Wahid had spoken at last. His accent was different to his wife's. Hers was East London. He told me that the problem had begun when he and his family lived in Somalia, where he was born.

"I was twelve ...," Wahid started.

It was nighttime. Wahid was asleep in the room he shared with his two younger brothers. His parents were asleep in the room next door. Suddenly Wahid found himself awake, sitting up in bed with his brothers staring at him. Before Wahid could work out what was going on his parents came running into the room. The younger boys had apparently called out and woken them. Wahid had been vaguely aware of a lot of noise and fuss but couldn't quite make sense of it.

"What's the problem?" Wahid's father asked the boys after he had turned on the light in their room.

Wahid's brothers didn't seem to be able to say what had upset them. They just babbled incoherently. Wahid had apparently woken them, but neither of the two other boys was old enough to explain clearly what he had done. Wahid was no help either. He found it all very confusing. He knew something odd had happened but couldn't explain it so he chose to tell his parents that he had no idea why his brothers had called out.

"Do you know if you were making sense when your parents came into the room?" I asked.

"I was fine. That was the thing. I thought they were all going crazy."

Finding each of their sons apparently well, the parents just remonstrated with them for waking the household and told them all to go back to sleep.

Thirty minutes later identical shouts brought the parents running again, with very similar results. This time the brothers reported that Wahid had looked as if he was scared and had kept pointing at something in the corner of the room. They thought he had seen something there, but they didn't know what. Wahid denied it. His father gave a cursory look around the room to try and understand what was frightening the children. Ultimately the boys got another telling off and a threat of punishment if they didn't go straight back to sleep. The rest of the night passed peacefully.

The next morning, with everybody in a hurry to get ready for school or work, nothing further was said about the incident. In fact the whole episode was completely forgotten until exactly two weeks later when it happened again. It was the same as the first time. The boys had been in bed for an hour. The parents were in the kitchen when the youngest boy came trotting in. He told them that Wahid had woken them up again. The brothers had heard him grunting and found him sitting up in bed pointing at the wall. When questioned Wahid continued to deny that anything unusual was going on. The parents didn't know what to do so they told all three boys to stop getting out of bed at night and left it at that.

Over the next four months Wahid's brothers intermittently complained that he kept grunting and pointing like a zombie at nighttime. When his parents tried to interrogate Wahid about this he became upset.

"I was always in trouble even though it was not my fault," Wahid told me.

Starting to become concerned, his mother took him to the doctor. He found Wahid healthy and suggested that he was suffering from nightmares. Advice was given about his diet and sleeping habits. Wahid's mother made the adjustments suggested. They didn't help. The problem just got worse and began to happen every week.

Eventually the parents decided to separate the boys at night. Wahid's father was consigned to the children's room and Wahid moved in to sleep with his mother. On the third night of sleeping together his mother was woken by a noise. The bed rocked gently. She turned to find her eldest son sitting upright with his head turned so that he was looking over his left shoulder. His left arm was outstretched and he was pointing with his index finger towards a spot on the wall. His mother thought he looked frightened. She looked at the wall and saw nothing. By the time she looked back at her son he was sitting in a relaxed position with quite a normal expression on his face. When she asked him if he had seen something in the room he said he didn't think so.

"So what were you pointing at?" she had asked.

"I don't know," he replied.

"Did you know you were doing it?"

"I'm not sure," he said.

"You must know what you were pointing at?"

"No."

Twice more that week Wahid's mother was woken by her son. She decided to consult her doctor again. He still insisted that Wahid was having nightmares. When the family rejected that diagnosis the doctor suggested that Wahid was just looking for attention. The parents were frustrated. They changed Wahid's diet again. They changed his mealtimes. They made him go to bed earlier and, when that didn't work, later. They asked the school if he was struggling with his work but they hadn't noticed any problem.

Feeling helpless, the parents decided to get the opinion of a traditional healer.

They gave the healer a detailed description of the events. Waking at night. Pointing and staring into the corner of the room and then denying doing so.

"Which wall does he point to?" the healer asked.

"The north wall, I believe," the father said.

"North to where Anwar's house is?"

"Indeed."

The healer said he knew the source of the problem.

"I have seen this one before!" he told them. "I know this one!"

The opinion he offered made a certain sort of sense to the family. A spirit was haunting Wahid. It was deliberately waking him at night with the intent of disturbing the family. The healer guessed that Wahid was fixated by the visitor and when he saw him he was compelled to follow him avidly around the room. By pointing Wahid was trying to warn others about what he could see. That the spirit was visiting with increasing frequency meant that Wahid had a message that he had not yet delivered. Or had not yet understood.

"What message?" the parents asked.

"He seems a very discontented spirit," the healer had said quite ominously. "I think it is Anwar himself. The grandfather."

Wahid's grandfather had died five years before. It was widely known that a family feud had erupted over the ownership of some of the grandfather's land, with Wahid's father and uncles fighting over their share. As the eldest, Wahid's father had taken possession of the land. His brother thought it should have been divided equally amongst all the sons. The healer was certain that this conflict was the reason for the haunting. That Wahid's accusing finger pointed in the general direction of the disputed land only served to strengthen his conviction. What's more, Wahid was the eldest. Any land taken in bad faith by his father would ultimately pass to him. The healer's interpretation fit neatly into the family's fears and guilt. In that moment it proved easy for them to accept. The healer suggested that they give part of the land to the brother who had originally laid claim to it. If they did so, a wrong would be put right and the spirits would be appeased. Reluctant at first the family eventually did as they were advised. It didn't work. Wahid got no better.

On learning of the failure of the treatment the healer brought in the local priest for his advice. They agreed that the grand-father must still be angry at the slight. More reparation was obviously required. They advised a further donation be made. This time a live goat was given to the church and another to the healer. Wahid's condition remained unchanged.

Over two years things continued much like this. Wahid's family followed the instructions of the healer and priest. They administered a variety of herbal remedies drawn from local plants. They prayed to the appointed gods. They sacrificed a chicken. They did each using the very specific techniques taught to them. It was all to no avail. Only when the family could not afford to pay for any further treatment and had no amends left to make did they give up and accept that Wahid would have to live with things as they were.

The attacks continued. Wahid and his brothers just learned to ignore them. By the time he was twenty-one Wahid was waking every night, three times per night. He came to accept this as part of his life. That it happened exclusively at night had the advantage that it made it very easy for the family to ignore. In the daytime he was quite well.

It was not clear where exactly this would have led had Wahid's life not undergone a major change. Wahid had studied economics in Somalia and when he was twenty-three he was accepted on a master's course at a college in London, where he went to live with an uncle whose own children had grown up and moved out.

The move to a new country required a big adjustment, but he coped. He discovered that he liked living in England – apart from the weather and being so far away from his family and friends. But he made new friends, at college and through his local community. The friend who would make the greatest difference to his life was Selma.

Selma had lived in London all her life. She worked as a hospital receptionist. Wahid and Selma met when his aunt invited Selma and her parents to dinner. Selma's family came from the same area of Somalia as Wahid. The pair liked each other immediately. It was with the joy and approval of both families that they became engaged ten months later.

Until Wahid married and moved in with Selma he had spent every night in London sleeping alone in his cousin's old bedroom. Nobody in his new home knew what happened at night. It was not so much a secret as of no importance to Wahid. So Selma didn't know. She learned about the attacks on the first night the couple spent together.

Selma had not yet fallen fully asleep when she felt her new husband sit bolt upright in bed beside her. She felt him moving around but didn't turn to look at him. Because it was their first night together she thought his sleep must have been disturbed by anxiety. Wahid didn't bother to explain himself and she didn't ask him to. It was only when the same thing happened the next night and the night after that Selma began to worry. She tried to ask Wahid what was wrong. He told her it was nothing. The more it happened the more Selma pressed him to discuss it. Wahid was very reluctant. Only when she persisted did he tell her the full story. When she learned what the traditional healer had said she took it badly. She told her new husband that he must be a fool if he actually believed in ghosts and spirits.

"He took the story back when I said what I thought," she told me. "But he still wouldn't see the doctor."

In fact it was several months later that he finally agreed to discuss the problem with his GP. A few weeks after that, we met.

"I've had this all my life. It's fine," Wahid threw his hands in the air in frustration.

"It sometimes happens four times a night," Selma said, and then, turning to her husband, "and it's not normal and I just want to know what it is."

"Are you aware that it's happening, Wahid?" I asked.

"Yes. I know I'm doing it but I can't stop it."

"What does it feel like?" I asked.

"I can't breathe. Like my throat is closing over. My whole body is stiff."

"Do you know what's going on around you? If your wife talks to you, can you hear her?"

"I hear her. I wait for it to be over to answer."

"Is it frightening?"

"Frightening?" He thought for a moment, "Maybe. Most of all it is painful."

"Painful?"

"Yes, my muscles, they hurt me."

"I have a video of it, if that will help," Selma said, taking her mobile phone out of her handbag.

"Brilliant."

Selma's and Wahid's descriptions had been so clear that I already had a very strong sense of these strange events. But a video was always welcome. Eyewitness accounts are very unreliable. Whether they are of car accidents or crimes or medical emergencies, accounts provided by onlookers are rife with error. Our brains are more sophisticated than any supercomputer, but they do not record events in a reliable and reproducible way like a computer. People imagine things that they didn't see based on their expectations. They miss other things. When we focus on one object our brains very easily and sneakily filter out details on the periphery of our attention. When questioned about events after they have happened, even the style of questioning influences the answer.

Selma pressed play on her phone and passed it to me. The video started and I could see Wahid lying in bed. The room was brightly lit. He had the duvet pulled tightly around himself. Only the top of his head was visible.

"It's about to start, you'll see him properly then," Selma said.

She had barely finished speaking when I saw Wahid in the video sit up in bed. It was quite sudden. He did seem frightened. I kept watching and listening. The Wahid sitting with me shuffled and looked in the opposite direction to the phone.

"He doesn't want to see it," Selma explained. "He didn't want me to video it."

I turned up the volume on the phone and heard a distinct gulping sound. Then a grunt. The sound that had woken the brothers. The Wahid sitting opposite put his hands over his ears. The one on the screen made another sound. An indistinct guttural sound. The picture was blurred but I thought I could see Wahid repeatedly swallowing. His eyes were wide open and they began to move. They tracked gradually to the side, very much as if he was watching something traveling slowly around the room. His eyes didn't stop roving until they were looking all the way to the left. Only the whites were clearly visible. Then his head followed in the direction that his eyes had taken, until his neck was stretched and his head had gone as far as it could. At the exact same time that his head moved, his left arm rose upwards and outwards until it was perpendicular to his side. His index finger pointed severely just as I had heard described. It did indeed look exactly as if he had seen something that nobody else could see.

"Well done videoing it from the very start," I said to Selma.

Most videos of strange attacks begin in the middle. It is hard for a witness to film them from their very first second.

"It was easy. He goes to bed and in the first two hours it happens. I just had to wait. I left the light on too," Selma told me.

"You were awake for it?" I asked Wahid.

"Awake enough. I couldn't talk, but I knew Selma was filming me."

Wahid was still staring away from Selma and me as he spoke.

"Do you know what it is?" Selma asked.

I did know. Much medical diagnosis relies on recognizing a familiar story when you hear it. I had heard stories like Wahid's before. I had also seen videos like this before. Many of them.

Brief attacks. Waking a person from sleep several times per night. The same every time. Forced head turning. One arm stiffening. Swallowing. Choking.

I wondered if they already knew what I was going to say.

"I don't know if this has been mentioned to you before but everything you've told me and what I've seen in this video are all entirely in keeping with the probability that you have epilepsy," I told Wahid.

He didn't immediately react. Selma's phone lay on the table in front of me with Wahid's frozen image on the screen. After a moment he reached over and turned it face down on the table.

"I don't want to see that," he said.

"Epilepsy?" Selma said.

Wahid and Selma looked at each other. Neither was convinced. They spoke hurriedly in their own language. The word *epilepsy* recurred a couple of times on either side.

"But he's never had a seizure," Selma countered.

"What I'm trying to say is that I am quite sure that these attacks are seizures," I pointed to the phone.

"They only happen when he's asleep. I spoke to a doctor in the hospital where I work and they wondered if it is a kind of sleep walking," Selma told me.

"It would be very unusual for sleep walking to happen every single night and several times per night. There are lots of different sorts of seizures and I am convinced this is one sort."

"I'm awake for it," Wahid countered.

"Yes, I understand that. Not everybody with epilepsy loses consciousness when they have a seizure."

"Just pointing can be a seizure, even if you know you are doing it?" Wahid asked.

"Yes, it can."

"I think I would prefer him to have tests rather than just guessing," Selma concluded.

I wasn't guessing, of course, but if it helped Wahid and Selma to have proof I would look for it. I arranged for Wahid to have a brain scan. The results revealed nothing abnormal.

MRI scans give us pictures of the brain that are very detailed. They allow us to detect tiny scars, tumors and blood vessel abnormalities that a CAT scan could not. However, there are still many medical conditions they cannot detect. MRI shows us the integrity of the gross anatomy of the nervous system – the brain's hardwiring – but brain disorders do not always affect the solid structure. They may only exist at a chemical, microscopic or electrical level. I suspected that Wahid had epilepsy, that is, a disorder of electrical activity. The brain may be structurally entirely sound in epilepsy. An MRI scan is blind in the face of an unwanted electrical surge. Fortunately, MRI and CAT scans are not the only ways to interrogate the brain. Medical investigations need to be versatile enough to allow us to look at the brain in lots of different ways to reflect the many different pathological processes at play. As epilepsy is a disease caused by spreading surges of electrical discharges, that is where we look next if the scan is normal.

The brain has its own intrinsic electrical rhythms. They are constantly changing. They drift and alter their appearance from waking to drowsiness to sleep. But they are also comparable and reproducible between people. In 1924 Hans Berger, a German psychiatrist, discovered that the brain's rhythms can be recorded directly from the scalp. Tiny as they are, brainwaves are measurable through bone and skin and hair.

Berger developed a simple technique to measure this activity and he called it an electroencephalogram – EEG. The brainwaves measured by the EEG have many interesting and clinically useful characteristics. Their patterns are different in different areas of the brain. The front of the brain and the back look electrically different. The patterns vary throughout the day and night to reflect a person's state of awareness. EEG recordings of the conscious and unconscious brain look nothing alike.

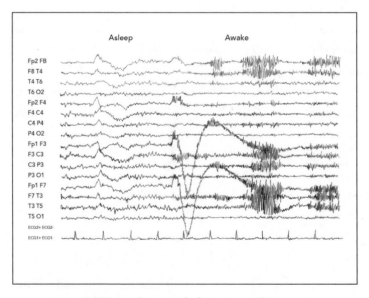

EEG in sleep and then on waking

Epilepsy can produce abnormal EEG phenomena such as spike discharges caused by a synchronous burst of electricity originating in electrically unstable brain cells. It is a marker for epilepsy. That is what I looked for next in Wahid.

He came to the hospital one morning and had an EEG. A technician applied twenty-five small metal electrodes to his scalp with paste. For thirty minutes Wahid lay still, even dozing off for a while, as the technician recorded brain activity from the surface of his scalp.

"Your MRI scan was normal, but the EEG is very suggestive of epilepsy," I told Wahid when we met again.

Wahid and Selma found it much easier to consider the diagnosis of epilepsy once they learned that the EEG was abnormal.

"Nothing to do with dead grandfathers then!" Selma said when I told them.

"No, nothing. But those sort of superstitions aren't as unusual as you might think," I assured them. In the 1980s witchcraft was still the second most often quoted cause for epilepsy in Nigeria. In 1999 a medical report described five cases of epilepsy in Florida, each of which had been attributed to voodoo spirit possession. Exorcisms for epilepsy are still performed in the UK.

Wahid agreed to take a drug for epilepsy. His seizures disappeared almost instantly.

*

Only six times in a thirty-minute EEG recording had we recorded a spike discharge in a discrete area of Wahid's brain. Like sparks from a fire, EEG spikes aren't there every time you look. You watch and wait for them. In total the six spikes did not amount to even a second of abnormality in thirty minutes. Their intermittent nature makes them easy to miss. A normal EEG does not mean a person does not have epilepsy.

The location from which a spike is recorded on the scalp is particularly important because that is what helps indicate where in the brain the seizures are arising. In Wahid the spikes were confined to his right frontal lobe. These microsecond bursts of brain activity were causing him no immediate problem, but they indicated a group of brain cells firing synchronously outside his control. Every night when Wahid went to sleep an electrical short circuit must have been spilling over to adjacent brain cells, resulting in a seizure. The epilepsy drugs Wahid took would not cure his problem, but they would reduce the electrical excitability of his brain, thus helping to prevent any stray sparks from spreading to become a seizure.

That Wahid's seizures do not look like the more traditional view of a seizure – a convulsion – is explained by the anatomy of the brain and understanding how an epileptic seizure develops. Seizures occur when an abnormal and autonomous electrical discharge transiently takes over part of, or all of, the brain. That discharge can happen for a multitude of reasons – head injury, birth injury, infections, autoimmune and genetic diseases amongst them. But ultimately every seizure stems from our neurons.

Neurons are the functional cell of the cerebrum. They have a cell body from which extend dendrites and an axon, which are like long arms reaching out into the center of the brain. Neurons are electrically active. They communicate with each other through the connections made by dendrites and axons. Each neuron has connections with hundreds or thousands of others. Everything we do is dictated by these connections. An EEG measures the synaptic transmissions between neurons.

Neurons are only 0.01–0.05 mm wide. Discovering their existence was a challenge because it required more than the development of the microscope. Blood cells can be smeared on a slide and put under a microscope very easily. The brain, however, is a solid, rubbery organ. To appreciate its microstructure scientists had to learn how to slice it extremely thinly. In the nineteenth century they learned that they could harden the brain by immersing it in formaldehyde, allowing it to be sliced without disturbing its architecture. When that was done the microscopic appearance of the brain was revealed for the first time, and with it a secret – the apparently homogeneous cerebrum was not so homogeneous after all. Different areas had different arrangements of cells.

At the very start of the twentieth century a German neurologist called Korbinian Brodmann became convinced that the arrangement of neurons said something about brain function. He painstakingly drew up a map of the brain dividing the cortex into fifty-two sections according to the histological appearance of each section. He referred to the mapped regions as Brodmann areas and designated a number to each. Work in neurostimulation would ultimately prove that Brodmann's artificial division of the brain into functional areas was largely correct. Our brains are designed according to a plan, and that plan is the same in each of us. The Brodmann area in which a seizure starts is crucial in determining what exactly will happen in that seizure.

There are many different sorts of seizures, but a basic classification sees them broken down into two main types – generalized and focal seizures. The traditional view of a seizure, a convulsion, is a generalized seizure or generalized "tonic clonic" seizure (tonic means stiffening, clonic means jerking). In a generalized seizure there is prominent tonic stiffening and rhythmic clonic jerking of the whole body. The word *generalized* refers to the fact that the whole cortex is involved in the synchronous electrical discharge that causes the attack.

Some generalized seizures are *primary* generalized, meaning that the unwanted electrical activity engulfs the whole brain simultaneously from the start. Some are *secondary* generalized seizures, indicating that the discharge has begun at one point in the brain and has spread to involve the whole cortex. The ends of a primary and secondary generalized seizure look exactly the same – tonic clonic jerking of all four limbs with profound loss of consciousness. The difference between the two is at the beginning. A primary generalized seizure is abrupt in onset and involves the whole body from the start. A secondary generalized seizure begins more gradually, often with a march of symptoms – a hand jerks first, then an arm, then one side of the face and so on. These spreading symptoms reflect the movement of the electrical discharge through the brain. The initial symptoms of a secondary generalized seizure are the clues that a neurologist looks out for to chase the seizure back to its point of onset in the cortex.

Not all generalized seizures are convulsions. Another sort is an absence seizure – a momentary blank spell that usually occurs in small children. Another is a myoclonic jerk – a lightning-quick jerk of the body. What all generalized seizures have in common is that the discharge involves the whole cortex.

Focal seizures are distinctly different. They are not so indiscriminate and all-encompassing as to take over the whole brain. They start in a small cluster of neurons in a single area of the cortex. From there the discharge may spread a little or a lot. The number of possible symptoms of a focal seizure is much larger than in a generalized seizure – different bits of cortex do different things and therefore produce very different symptoms. Some focal seizures will become generalized, but many stay focal, only ever affecting a very circumscribed area of the brain. The average brain has in the region of 85 billion neurons. Focal seizures could begin in as few as 2,000 of those. Which few matters a great deal.

*

Cherylin, another of my patients, also has epilepsy, but her experience of it has been very different to that of Wahid. Her mother had described Cherylin's seizures to me. "She goes away with the fairies, mad as a hatter," she said. Cherylin knew nothing about what happened when she had an attack. It must be so strange to have a disease in which the only person who is not really present for its appearance is you; the person who knows the least about it, you. Cherylin woke in strange places with no idea how she got there. She might as well have been teleported. Waking scared and disorientated were all that her seizures were to her.

"I come round with the awful feeling that I'm about to die," she told me.

Cherylin liked to have family with her when she woke up from a seizure. She needed the assurance of somebody she could trust. I felt myself a poor substitute the day that I was the only one there to help. Cherylin had come to clinic unaccompanied. This was unusual, but few of us have the luxury of a loved one who is always available, and on this particular day Cherylin's family were all busy.

She was just in the course of telling me how she had been since we last met when her whole expression changed. Something passed across her face. Asked to describe it I would struggle. An infinitesimal pause. It was subtle so I wasn't sure. I dismissed it and continued talking.

"What dose of lamotrigine are you currently taking?" I asked her.

She didn't answer. She looked down at her hands, examined her fingernails. With her right hand she twisted the ring she wore on her left thumb. After a few seconds she looked up at me again but still didn't answer. Maybe she hadn't heard the question? Was she unsure of the dose or was it something else?

"It says in your notes you take a hundred milligrams – is that right?"

This time I was more certain that something was wrong. Contact was completely lost. Abruptly she began to count. Loud frightened words.

She was having a seizure.

"One, two three, four, five, six, seven, eight, nine, ten, eleven, twelve, thirteen …" she shouted. As the numbers got higher so did the volume and with it a sense of fear. Her brain was working at hyperspeed, on full display. In response my own brain felt sluggish. Seizures are shocking. They are sudden. Flagrant. They take away the control of the person affected, but also that of onlookers. When the brain catches fire like that there is very little an onlooker can do.

I formulated my response. I was too slow. Before I could properly react Cherylin stood up from her chair and shuffled backwards into the corner of the room. Still counting.

"Twentyseventwentyeighttwentynine …"

She looked as if she had seen something so awful that she needed to make herself as small and invisible as possible. I moved towards her, muttering nervous placating sounds.

"It's all right, it's all right, I'm here. You're safe," I said.

I hunched down beside her. She grimaced, took my hand and squeezed it tightly. With her back pressed against the wall and her knees bent she reminded me of a small frightened animal.

"Oh no, oh no, oh no," she said breathlessly, and then, "help me, help me. You'll help me, won't you?!"

Her pleas only heightened how useless I felt. She needed reassurance for a fear that existed beyond my reach. It lived only inside her brain. I put my free hand on her shoulder as if that would be enough to soothe her, but of course it was not. The screaming became more urgent.

"Help, help me please, help me please …"

She was still holding on to me very tightly. I knew I would have to pry her hand from mine if I was to reach the emergency alarm button that signaled the need for help. I regretted that

that was not the first thing I had done. She burrowed as deeply as she could into the corner and pulled the neck of her jumper up over her mouth and nose. Her eyes peeped out, roving around the room.

When a patient has a convulsive seizure the correct reaction is to roll them on their side and wait for it to end. Lying on your side protects you from choking. That was not practical in Cherylin's sort of seizure. Her back was resolutely pinned to the wall. She was not going to choke, but she looked poised to spring. If she did it was my job to keep her away from anything dangerous in the room. That was all that was within my power to do.

I also needed to make sure that the seizure passed in a reasonable amount of time. Most will end in a couple of minutes. Longer than five minutes is a potential emergency. I looked at my watch. I guessed less than a minute had passed. I began timing it from there.

"No, no, no, help, help, help me ..." Cherylin's words were muffled by the wool held tightly against her mouth. I tried to pull her jumper down from her face. She looked right at me, widened her eyes and shook her head defiantly.

I looked at my watch again. Another thirty seconds. It had only been a minute and a half but it felt much longer. I wished I could free my hand from Cherylin's and go to the door, but to pull away from her seemed cruel. I began to worry that the seizure would never end. Soon I would have no choice but to force her fingers open to go for help. At the last second I was saved. There it was again – a subtle change in the expression in her eyes. Back to normal, or near normal at least. Cherylin's grip on me weakened. She was still crouched on the floor but the specter had lifted. She uncovered her face.

"Are you okay, Cherylin?" I asked.

"Yes."

"You're back with me?"

"Yes."

"Do you know what happened?"

"Is my mum here?"

"No. You came on your own today. Would you like me to call her for you?"

"Is Mum outside?"

"No. Are you okay?"

She stood up and moved to where her bag lay on the floor beside her chair. A stain on her jeans told me she had wet herself. She picked up her bag and began to walk towards the door.

"Sorry," she said.

"Don't go," I said touching her arm in an attempt to direct her back to the chair, "sit down and let me get help."

"I'm fine."

"You can't go like this. You're wet." I indicated the stain on her trousers.

She looked down and rubbed the stain with a hand.

"I think I'll go home," she said.

It felt like we were involved in two separate conversations.

"Do you know where you are, Cherylin?"

"Yes," she smiled.

"Where?"

"I'm here," she smiled and gestured around the room with her hands.

"Where's 'here,' Cherylin?"

"The shops."

"No Cherylin, you're in the hospital."

"Can I go?"

"Not yet."

I propelled her to sit. I picked my pen off the desk and showed it to her and asked, "What's this called?"

"A pen," she said arching her eyebrows quizzically.

"What day is it?" I asked.

"What day is it?"

"Yes, what day is it?"

"Sunday?"

"No, it's Monday. Do you know who I am?"

"Is my mum here?"

"No. Do you know that you're in the hospital?"

"No? Am I?"

It took ten further minutes for Cherylin's answers to make absolute sense. Once they did she also became distraught. She cried and cried and begged for her mother to come. This was the feeling of doom she had told me she experienced as soon as any seizure was over.

I picked up the phone and called the epilepsy nurse to come and help. She took Cherylin to the coffee shop and brought her a cup of tea. Something to calm her as she familiarized herself with the world once again. Only when Cherylin seemed fully recovered did she come back so I could try to restart our consultation. It didn't work. Cherylin had lost the will to take part in the conversation. She just wanted to go home. We called a taxi and I arranged to meet her again the following week.

Before she left I asked, "What happens, Cherylin, if you have a seizure in the street?"

"Nothing really. I just wake up and people are staring at me and I walk away as purposefully as I can. I know I'll never see them again and it makes me feel better to think of it that way."

*

Wahid and Cherylin have the same diagnosis. How it affects them has been barely comparable. They both have focal seizures, but coming from a different area of the brain.

Cherylin's birth had been very difficult. Her mother's labor was prolonged and Cherylin was blue and floppy when she first appeared. She was rushed to the special-care baby unit where she spent almost a week. In total she spent the first two weeks of her life in hospital.

From the moment they took her home Cherylin's parents noticed a clear difference between her and their other children. Everything new she learned, she learned very slowly. Once she was in school their suspicions were confirmed. She underwent a formal assessment and was discovered to have a mild learning difficulty. It wasn't devastating news. She was a happy child who was well enough to attend mainstream school.

More upsetting was when Cherylin developed epilepsy at the age of seven. The scan showed scarring in her brain sustained during birth. Her seizures proved difficult to treat and twenty-five years later when she and I met they still happened regularly.

In contrast Wahid had a normal scan. He had no learning problems. His seizures looked nothing at all like Cherylin's.

They were brief and only happened at night. They went into remission on treatment.

Wahid's and Cherylin's seizures, whilst both focal, did not look at all alike because they each started in a different lobe of the brain. Wahid's seizures came from the right frontal lobe, whereas Cherylin's EEG showed spike discharges in the right temporal area and therefore were temporal lobe seizures. Throughout the twentieth century lesional studies, animal studies and neurostimulation slowly uncovered the gross functions of the four lobes of the brain. It is in the functional anatomy of the various lobes that Wahid's and Cherylin's differences can be explained.

Wahid's seizures manifested with predominantly motor symptoms – his head turned, his eyes tracked around the room and he raised one arm. The frontal lobes contain several regions central to the planning of and control of movement. Prominent jerking or stiffening of muscles are early symptoms of seizures originating in the motor regions of the frontal lobes. The frontal lobes also contain the frontal eye fields. These play a role in visual attention and allow our eyes to follow moving objects. They are the reason that frontal lobe seizures often involve tracking head and eye movements.

The predominant features of Cherylin's seizures were emotional. The temporal lobes contain areas pertinent to emotional control and expression. Cherylin's seizures were dominated by fear, terror, a sense of doom. These are all common manifestations of temporal lobe epilepsy and are reflections of what a healthy temporal lobe does for us. Although epilepsy is often incorrectly thought to imply the presence of convulsions, neither Wahid or Cherlyin have ever collapsed or had a generalized seizure. The electrical discharge has always stopped before it could involve the whole brain.

Tests were never really necessary to tell me what was wrong with Wahid or Cherylin. They were luxuries that allowed us all to have faith in a diagnosis that was plainly available in their stories from the start. Just knowing how the brain is organized and the manner in which electrical stimulation spreads from cell to cell was enough to explain their symptoms.

Epilepsy is the ultimate disease chameleon. Each one of 85 billion neurons could have hundreds or thousands of connections with other neurons. A typical brain has a 100 trillion synapses or connections. Focal epileptic seizures arise in small groups of neurons and spread through any or all of the selection of synapses available to them. This creates endless possibilities. Focal seizures arising in different lobes look different. Seizures arising in different parts of the same lobes look different too.

Every action we perform, everything we feel, every dream we have is a biological process that begins or ends in the brain, the result of movement of ions in and out of cells, the flow of an electrical discharge, the release of chemicals. Those processes are the same in each of us but we are not the same. It is the endless variables within the circuitry of the brain that ensures we are all unique.

2

AMY

Either the well was very deep, or she fell very slowly, for she had plenty of time as she went down to look about her and wonder what was going to happen next.

—Lewis Carroll, *Alice's Adventures in Wonderland* (1865)

The seeds of epilepsy were sown in Amy's brain when she was only three months old. It took sixteen years for them to be revealed.

She was a healthy baby. A second child, her parents remembered feeling altogether more relaxed as parents than they had with her older sister. Her mother brought pictures of Amy as a baby to our first consultation.

"I want you to see what a beautiful baby she was," she said as she spread out on the desk in front of me a set of photos of an apple-cheeked smiling baby. "And this is her in hospital," she said taking out a second set of photographs, which showed the same child but now unrecognizable, lying in a hospital bed with tubes coming from her mouth and nose and arms.

"How long was she in hospital for?" I asked.

"Two days in the intensive care unit and then two more weeks on the ward."

Amy had become sick and feverish while on a family holiday. Since they weren't at home her parents had prevaricated slightly longer than usual over whether or not they should consult a doctor. With incredible speed Amy's temperature rose and she developed a mottled rash on her abdomen. Her parents became worried when they saw that and decided to take her to the local hospital. On the drive there the rash became more florid and started to cover her whole body. Carrying her into the hospital her mother noticed that some of the dots on her skin had turned black. She was floppy and lifeless.

"She was like a rag doll," her mother told me.

Amy was very quickly diagnosed with meningitis and received emergency treatment. She had a seizure in the intensive care unit and was placed on a ventilator. The family prepared for the worst but it didn't happen. Amy responded to antibiotics and slowly recovered. The blackened spots on her skin receded. In a week she was eating normally and smiling. By the time she went home there were only small traces of residual scarring on her skin, which ultimately disappeared.

It seemed that Amy had escaped unscathed. Her parents had worried for a while that her hearing was impaired. Deafness is a frequent complication of meningitis and she appeared less attentive to noises in the months after the illness. That soon resolved. By her first birthday there was no evidence of her hospital stay.

Then, when she was sixteen, Amy developed epilepsy. She had a convulsion, a generalized tonic clonic seizure, at school. She didn't receive treatment after the first attack. While five to ten percent of the population will have a seizure at some point in their lifetime, only some of those will go on to have a second seizure, and epilepsy is defined as recurrent seizures. The approx-imate average risk of having a second seizure is just under fifty percent, so the usual medical strategy after a first seizure is to watch and wait. Injuries to the developing infant brain, what-ever the cause, can lead to epilepsy later in life. Once Amy told her doctors that she had meningitis as a child they would have known that she was at high risk of having another seizure, but still they avoided treating her just in case. Amy didn't have to wait long. She had a second similar seizure a month after the first. She was diagnosed with epilepsy and started on treatment.

"Describe the seizures to me," I said to Amy when we first met.

At that point Amy was in her late twenties. Since her diagnosis she had had good and bad years. As a teenager she had seizures every month. At the age of twenty-one they disappeared for eighteen months. In the UK people with epilepsy can drive if they have had no seizures for one year. Amy had just started taking driving lessons when the seizures came back. After that they came and went, never staying away quite long enough for her to drive again. However, in every other sense Amy's life moved forward as normal. She went to university, where she studied marketing. I met her when she moved to London to work for an advertising company.

The first time I meet a patient I need to hear a detailed description of the attacks. To open the window into the brain.

"I have two sorts. I call the main type my Alice in Wonderland attacks," she told me.

I am fascinated listening to people describing their seizures. Most offer very personalized descriptions that encapsulate the experience for them: *meltdowns, hokey-cokeys, electric shocks, screams, heil Hitlers, headbangers*. When I record a patient's seizures in their notes I always include these names. They are more vivid than any medical terminology.

Seizures are usually a very frightening, disruptive experience for patients. Often they are very unpleasant. But not always. Every now and again I meet somebody who feels a certain privilege. Some say it gives them a unique perspective on the world. Not all the time, but sometimes, they get to view a world that nobody can share.

Amy didn't love her seizures but she did love the feeling she had when a seizure was about to start.

"If I could have the aura at least once every day, I would," she told me.

Many focal seizures begin with what is referred to as an aura, from the Greek word for *breeze*. It was first used in the context of epilepsy in about AD 200. A young boy describing his seizure said that it began with a feeling like a soft breeze blowing on his leg. The aura is a manifestation of a focal electrical discharge beginning in a single spot in the brain. If the discharge stops at that point, nothing further will happen. If it spreads then the symptoms will evolve with it. An aura is a warning. The term is no longer used only to refer to a sensation like a breeze; it is any transient symptom that happens at the beginning of a focal seizure. Common examples are déjà vu, butterflies in the stomach or hallucinatory smells. To a neurologist the aura is the first clue.

Amy liked her auras, but not her seizures.

"After the aura it becomes more frightening?" I asked.

"Yes. The next bit is a drag."

Amy's seizure began with a sense of disorientation that she quite enjoyed. She likened it to taking a pleasant hallucinogenic drug. Sometimes that was all her seizures were. An aura. A happy feeling that made the world look beautiful. Unfortunately not every seizure stopped there. The electrical discharge that began in a circumscribed area of her brain occasionally spread through her cortex until the whole brain was involved. When that happened, happiness gave way to a generalized convulsion. Unpleasant and dangerous, they put her in compromising positions. Sometimes she woke in the street surrounded by strangers. Once a friend found her lying half undressed on the bathroom floor. Nobody knew how long she had been there. A particularly frightening episode occurred in her own kitchen while she held a chopping knife in her hand. When she fell she landed face first on the blade and suffered a serious laceration. A raised purple scar under her right eye is a permanent reminder of that seizure.

Despite episodes like this, Amy remained cheerful.

"If you could give me a drug that gets rid of the big seizures but lets me keep the little ones I wouldn't say no," she told me, laughing.

"Keep going with the description. You feel pleasantly disorientated and then?"

"It's so hard to describe, you have to experience it yourself."

Seizures are a challenge to put into words.

"If you do your best to describe them, then I will return the favor by doing my best to only make you half better!" I joked.

"Okay," she hesitated thoughtfully, "I get a feeling that lets me know it's about to start. Now don't ask me to describe that bit. It's just a feeling, there are no words for it."

"Nice? Not nice?"

"Oh! *Really* nice. Really properly full-on nice. Like everything makes sense."

"I wish I had that feeling."

"It is good!"

"And then what happens?"

"So … that bit lasts a second or two. Although I can't be sure about that because time gets really tricky and hard to measure. Then I notice things changing around me. Whatever I'm looking at begins to move. The food on my plate. The television. They seem to sort of slide away from me and at the same time they seem to shrink. They're shrinking because they are getting further away – that's what I always think," she paused, "… although sometimes it's a bit different, and I can't decide if I'm getting bigger and everything else is getting smaller. The next thing is that the ground in front of me changes. It looks like it's sloping and getting thinner the further away it gets. Do you know what it's like?!" Amy suddenly pointed her finger in the air to indicate that a comparison had just struck her.

"What's it like?"

"Like a road in a painting. A really bad painting, you know? Where the road is getting thinner and thinner to show you that it's far away. But you know it isn't really a road and it isn't really far away."

"That's such a good description. I don't know why you said you can't describe it."

"I suppose because I know I'm not getting it quite right. I have to be in it to really explain it – but if I was in it I couldn't explain it."

"Well I think you're doing a pretty good job. Is there more to it?"

"They all start in that way. Most just sort of wear off then, but sometimes it gets worse and that's when I pass out."

"That's when they get frightening?"

"Yeh. I feel myself beginning to slide downhill – even if I'm on completely flat ground. If I'm walking when it happens I actually start to walk as if I was going down a very steep hill. I know there is no hill there but I walk like there is. Or at least I always thought I did but my mum says I don't. She says I'm walking normally."

"You're awake for all of this?"

"Yes. Until I start to get sucked down the road – then I pass out."

"So that's what you call the Alice in Wonderlands?"

"Yes."

*

The person best known for using neurostimulation to plot brain function was Wilder Penfield, an American neurosurgeon working in Canada in the mid-twentieth century. He used cortical stimulation as a way of systematically exploring the functional neuroanatomy of the cortex. He represented his findings as a diagram referred to as the cortical homunculus. The homunculus is a cartoon man shown languishing over the surface of the brain. Each of his body parts is used to indicate where the corresponding motor or sensory processing for that part happens. The proportions of the homunculus are grotesque. He has a giant thumb and tongue relative to a small torso. These size ratios represent how much more of the cortex is needed to serve the complex sensory and motor requirements of one body region compared with another.

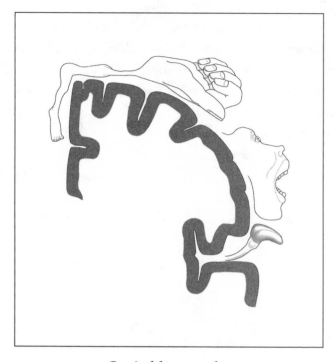

Cortical homunculus

Brodmann tried to understand the brain by creating a histological map. Penfield's map explored function. Despite their
limitations both were remarkably accurate. The homunculus
for the motor region roughly corresponds to Brodmann area 4.
The strip of cortex where sensation is processed relates reasonably closely to Brodmann areas 1, 2 and 3. But the problem
with these maps is that they threaten to be misleading by
suggesting that one area of the brain represents, and is responsible for, one function.

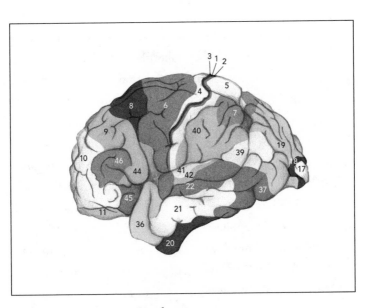

Brodmann areas

The way the brain processes visual information is a very good example of the challenge posed to the early brain explorers. None of the techniques available to them could provide a full understanding of the complexity of how the brain works. Functions could only be tested at their most basic level. Only those functions that were easy to observe, or measure or describe could be studied. Voluntary movement, sensation and speech were much simpler to assess than higher functions involving thought and emotion.

Penfield's neurostimulation correctly demonstrated that visual processing took place in the occipital lobe at the back of the head (Brodmann area 17). But the brain doesn't process vision as a camera does – as if the eyes take a picture and the occipital lobe records it. Visual stimuli are actually subject to several stages of processing and they take place in scattered regions throughout the brain. Neurostimulation could never have detected this.

It really wasn't until the twenty-first century that science finally developed the capability to understand the brain at a more sophisticated level. That came with the invention of functional MRI (fMRI). Standard MRI shows only anatomy; as the name suggests, fMRI is used to assess function. It uses a standard MRI scanner but applies a statistical analysis that compares blood flow in the brain when a person is asked to perform a task and when they are at rest. For example, a person undergoes MRI scanning alternatively listening to music and then white noise. The difference in the two pictures is used to infer which brain regions are involved in processing music.

Functional MRI has proven to be the most extraordinary tool for looking into the thinking brain in detail, but it does have limitations. MRI pictures are only shadows from which huge inferences must be made. Where there is deduction there will be error. A sobering study was carried out in 2009 that reminded us of this fact. A group of scientists conducted an fMRI-based study on a dead salmon. They showed the salmon pictures of humans in social situations and (comedically, one assumes) asked the fish what emotions were exhibited in each picture. Serial scanning was done during the questioning. When the resulting scans were compared they showed apparent areas of activation in the salmon brain during the questioning. These could easily have been interpreted as physiological, but (because that was impossible) they were actually a clear demonstration of a false positive effect attributed to statistical analysis. If enough statistical tests are done on any set of scans, some will, by chance, be positive.

Keeping these limitations in mind, though, fMRI has certainly added to our understanding of how the brain works. It has shown that visual processing is not restricted to Brodmann area 17. Understanding this goes some way at least towards helping explain Amy's strange Wonderland-like experiences.

When we look at something we see a unified object. But our brains don't process vision in a simple or straightforward way. The final image is a construct. Visual processing requires the effort of connected neural pathways, not all of which exist in the occipital cortex. The primary visual cortex in the occipital lobe only makes sense of what we see at the most basic level. It is the first stage of several. From there visual stimuli are subject to a hierarchy of processing in multiple brain regions. Much of the more detailed processing takes place outside the occipital lobes. The primary visual cortex has many connections. It very rapidly sends information to the parietal and temporal lobes. The temporal lobe is likely to be where the most sophisticated consideration of the visual signal takes place.

When we look at an object we need to judge its depth, color, shape. We assess the ambient light in the room. We decide if an object is moving and, if so, in what direction and how fast. We decide if the object is already familiar to us. Functional MRI has been used to help us know where in the brain all of this is done. One region of the temporal lobe seems to be important to the detection of linear and circular motion. A distinct region nearby, also in the temporal lobe, activates when we look at shapes and colors.

Because vision is processed through connections between different brain areas it is possible for one aspect of visual processing to be affected by a disease while the others are preserved. In Oliver Sacks's book *The Man Who Mistook His Wife for a Hat* he describes an artist who loses the ability to recognize faces. He could identify objects but not his own wife. Had MRI been available in that man's lifetime it is possible and even likely that it would have shown the man to have a lesion of the fusiform gyrus, an area of brain on the undersurface of the temporal lobe that is implicated in the recognition of familiar faces. (Interestingly the same region activates when a car enthusiast sees a car they admire. No brain region is functionally exclusive.)

Only as the brain is understood can the bizarre impact of disease on the brain be understood. Visual perception requires the coordinated effort of multiple brain regions. A seizure can pick off any stage of that process to different effect.

Jenna was a young woman with a curious complaint. She experienced intermittent fluctuating brightness in her vision. It lasted for only a minute or two but recurred several times per day. In between the experiences she could see normally but when it happened she struggled to focus. She never lost consciousness so she was able to give a clear description of what was happening as it happened. When doctors examined her while she had the symptoms they noticed something very unusual. Her pupils were contracting and dilating. In some seizures a limb jerks rhythmically but in Jenna it was her pupils.

This dilatation and contraction is referred to as hippus. Hippus is a rare manifestation of epilepsy. Jenna had an abnormality causing seizures at the junction between her right temporal and parietal lobes. Exactly how this caused hippus is open to conjecture. Given the location of the seizure in the brain the theory is that the seizure caused a transient disruption in Jenna's ability to process color, shade or movement. So perhaps her pupils were only responding to the confusing message they were receiving from the brain.

Amy's Alice in Wonderland seizures were visual illusions. An illusion is a misperception, a distortion of a real sensory experience – mistaking one shape for another. Amy's brain scan showed that she had suffered significant brain damage as a result of meningitis. There was extensive scarring in the right temporal and occipital lobes in particular. I played down the appearance of the scan when we discussed it.

"My brain is scarred?" she looked worried despite my reassurances.

"Yes, but it happened nearly thirty years ago! You've lived with it and done very well. It hasn't affected you very badly so don't let it now. The rest of your brain has compensated."

In truth I was surprised at how abnormal Amy's scan looked. I was surprised because Amy was so well, bright, a university graduate pursuing a challenging career. There was a disparity between her scan and her well-being. We used to think that the brain couldn't repair itself at all, that it was anatomically fixed, we never made new neurons and that loss of function was irreparable. We now know that's wrong, our brains show neuroplasticity – our wiring can potentially be remodeled and rerouted to learn new skills or compensate for damage. Children's brains have a bigger capacity for this cerebral reorganization than those of adults.

Perhaps neuroplasticity explained Amy's well-being, but even though the scarring had not affected her intellect it had caused her epilepsy. An EEG showed right temporal spikes. Her visual distortion was a manifestation of disruption of higher-level visual processing. During a seizure she could see objects but couldn't judge depth or perspective.

Unfortunately I was not able to stop Amy's seizures. The scans and EEGs only rarely contribute to making people with epilepsy better. They provide explanation more than cure. Thirty percent of people with focal seizures will never go into full remission. However, Amy's seizures did reduce in frequency and the tendency for the discharge to spread from a focal seizure to a more dangerous generalized one lessened with time. Amy kept her lighthearted view of the world throughout.

"I know lots of people have seizures much worse than mine. I need to stop complaining and remember how lucky I am," she told me once.

Amy had named her seizures herself in a way that meant something to her. She was shocked to learn that similar seizures are actually referred to as "Alice in Wonderland syndrome" by the medical community. Visual illusions reminiscent of Alice's journey are actually not that unusual. They are almost always associated with temporal lobe epilepsy. But the experience of objects looking smaller or bigger than they truly are is not exclusive to epilepsy; it is also associated with migraine.

There is even an unsubstantiated but recurring rumor that Lewis Carroll was an epilepsy sufferer. He certainly had migraines. Entries in his diaries recount his experience of headaches associated with the visual hallucinations of a zig-zag pattern which are typical of that diagnosis. But he also suffered two episodes of loss of consciousness that are less likely to be connected with migraines. To an epilepsy specialist his account of collapse, combined with the distorted world he brought to life in his books, makes a diagnosis of epilepsy very tempting indeed. It would be lovely to believe that Lewis Carroll transformed the awful chaos of recurrent epileptic seizures into something as beguiling as Alice. I hope he did.

3

DONAL

Every act of perception is to some degree an act of creation, and every act of memory is to some degree an act of imagination.

—Oliver Sacks, *Musicophilia: Tales of Music and the Brain* (2007)

The idea that the only reality we can ever know is our own has always fascinated and frustrated me. Every day I meet people who have experiences that are very unusual. Sometimes they are unique. I am called upon to determine if what they experience constitutes difference or disease.

I met Donal when I had been working as a consultant for only two or three years. A doctor's confidence in their own opinion is subject to highs and lows. As a newly qualified junior doctor you take a stance of feigned bravado backed up by whispered reassurances from senior colleagues and the calming rustle of the pages of a book. You act like a good doctor in the hope that one day you will feel like one too. You work long hours. When you are not working you're studying. The exams you do in the training years are more competitive than any you did as a medical student. They have the highest failure rate. This is the period when the most detailed and focused knowledge is accrued. It is not easy to pass the postgraduate exams. Some never do. Once you do pass them, you need only find your specialty and then take the first step on the steep ladder that leads to consultancy.

In time experience bolsters you. You become more certain of your own clinical judgment. It is just about then – in the final training year, or the first consultant years – that I fear you are the worst doctor you can be. *A little learning is a dangerous thing; there shallow draughts intoxicate the brain.* You are beginning to feel you have seen everything. But you haven't. You only become a safe doctor when you realise that you never will.

Donal came to clinic with his wife. I liked him from the outset. He was a stoic. Quiet. He spoke soberly, answering questions with little or no embellishment. He was somewhat impenetrable. He supplied me with facts, but not a lot more. When I asked him how many children he had, he told me three. He did not say if they were boys or girls, or if they still lived at home, or if they had children of their own. He did not add "they're great kids," or say "for my troubles" and then laugh as some people do. When I put a simple question to him, he answered it simply. That is more unusual than you would think. It interested me. It made me ask myself if he could be a man who did not or could not express his emotions. I remember wondering if that might not be the problem.

Donal worked as a janitor in a school. He had been in the same job for thirty years. His wife, the more effusive of the pair, told me that he had always been very satisfied with his work. He took great pride in doing it well.

"You will not find a weed in that schoolyard," she told me. "He does all the small maintenance jobs. That school had no need of a plumber or electrician, only once or twice in twenty years."

Knowing that Donal's job meant so much to him also influenced my early conclusions. Months before we met, Donal had had an odd experience in a very specific situation. He had received a summons to see the headmistress. This was unprecedented. Usually he worked alone, not in need of any direction. Rumblings in the school meant that Donal had a strong suspicion about why he might have received the call. Changes, cutbacks, redundancies were the talk of the staffroom. My ears pricked up at this. I pressed Donal to tell me more. In response he admitted reluctantly that as he had walked the length of the corridor to the headmistress's office he had felt uncharacteristically anxious.

"Our children went to that school," his wife said. "He's seen children from start to finish in that place and then welcomed them back with their own children." She paused. "I don't know what he'd do if he lost that job."

Donal was right to be worried. The news was not good. The school was being forced to cut its budget and the head had called him to warn him that she was reviewing his hours. Redundancy was not an impossibility. The job that Donal did could be done for less money these days, he was told.

"They said he was the expensive 'option,'" his wife told me. "Option! Would a contractor sweep the leaves off every one of those paths before school every day?" she asked, incensed. "They would not!"

Donal had been seated when the headmistress gave him the news. He had stood to leave the room when he had his first attack.

"I want you to go through it again, but more slowly this time. Give me a moment-by-moment account of what happened," I said.

Donal had already described his experience to me, but I had never heard anything like it before. When the brain lays out its clues, every detail counts. If I was to follow them to their source I would have to understand them.

"I was sat down. I thanked Mrs. Daly," he said, his wife rolling her eyes, "then I stood up and turned to leave the room."

"How soon after standing did you feel unwell?" I asked.

Donal thought for a moment.

"Two, three seconds, no more."

"Did you feel at all lightheaded or unwell or sick in any way?"

"No."

Standing suddenly can cause an abrupt fall in blood pressure. If the blood pressure doesn't recover quickly enough it momentarily deprives the brain of oxygen. That can cause a person to feel sick or to faint. In some of us it causes a head rush. I was wondering if that had contributed to Donal's problem. But he said he hadn't felt sick or dizzy.

"You were walking towards the door, what happened then?"

"There's a big plant pot in the corner of the room with one of those yucca plants," he told me. "I was about three feet away from it. That's where they came from. Clear as day coming right from behind the plant pot. They did a running dash past me and disappeared behind the filing cabinet."

"They ran from right to left?" I asked.

"Yes."

"And what did they look like?"

"About a foot tall ..."

"I mean what exactly did they look like?"

Donal looked unhappy to have to tell me the story all over again.

"They looked like the seven dwarfs – like I already said. Seven small brightly colored men. They moved from right to left, quite fast. I had more of an impression of them, I would say, than a clear look. I knew who they were without maybe seeing them in every detail. I couldn't say what they were wearing, if that's the sort of thing you want to know."

"Did you know they weren't real?"

"Of course. Well, they were real I mean ... but not ..." He took a breath and gathered himself. "What I'm saying is I really did see them ... but not that I think that cartoon characters are real."

"Oh – they were cartoon characters?!"

"Yes. Did you think I saw little people?"

"Sorry, yes ... well it's quite an unusual story, so I suppose I didn't really know what you saw. Why do you think you saw them, Donal?"

"At the time I thought it was some projection from outside. Like children playing a prank, maybe. You've got to understand they were very very real," he said, "but when I saw them the next time, I was at home, so that was that theory in tatters."

"Did you tell the headmistress what you'd seen? Did she notice anything wrong with you?"

"I didn't trouble her with it. I mean, I said to her, 'Did you see something just run across the room?' She just said she hoped it wasn't a mouse. I said I thought not and left it at that."

"He didn't tell me either," his wife said, "not straightaway."

The second time it happened, Donal was in his dining room building a model of a battleship. This was how Donal spent much of his leisure time. As soon as his wife told me that, it made a lot of sense to me. He had a quiet, careful way about him. His posture was slightly stooped. I could imagine him bent for hours over a table, concentrating on precision work.

The room was separated from the kitchen, where his wife was, by a sliding door which was half open. His wife heard the sound of something falling and then Donal grumbling under his breath. She went to see what had happened and found Donal picking model pieces up off the ground.

"I've never seen that man drop so much as one piece, never mind knock over the whole thing," she told me, "but still he didn't tell me why."

"You saw them again?" I asked.

"Yes."

"Exactly the same?"

"Yes."

"Why did you drop the model?" I asked. "Was there a problem with your hands?"

"It was just the surprise of it," he told me.

It was only on the third occasion that Donal finally told his wife. The couple went to bed at roughly the same time most days. One night Donal's wife was woken by him sitting up abruptly in bed and gripping her arm.

"He gets up twice every night to go to the toilet," she told me. "He usually makes some sort of racket when he goes, but he had never grabbed hold of me before."

"I'm seeing things," Donal had said, holding tightly to his wife's arm.

She had turned on the light and looked at him.

"He seemed scared, but that was all," she said.

This time, perhaps influenced by the darkness and the confusion of the sudden move from sleep to wakening, Donal did not find it so easy to shake off what he had seen. He asked his wife to check under the bed.

"I thought he was playing silly beggars," she said.

Donal wouldn't say why but he insisted that his wife lean over the side of the bed and look underneath. Her husband was not the sort of man prone to foolishness, so she did what he asked. Needless to say there was nothing there.

"Just the usual balls of fluff and a stray sock," his wife told me. "I was relieved. With the way he was going on I was terrified of what I'd see."

"I just had a really bad dream. Really real," he said and they both eventually shrugged it off and lay down again to go to sleep.

"He didn't tell you what the dream was about?" I asked, to which Donal's wife gave the first hearty laugh I had heard from her.

"Him," she cocked her thumb towards her husband, "talk about a dream!" And she laughed again.

It was only when the same thing happened a week later that Donal finally told her more.

"I saw the bloody things again," he said.

"What things?" she replied, turning on the light to find Donal sitting up in bed looking puzzled.

"I've had the same dream I had before," he said and reluctantly told his wife exactly what he had seen.

"Seven fairy-tale dwarfs!" his wife recounted to me. "Can you believe it? I told him he was dreaming. That's when he admitted he'd seen them at the school in broad daylight."

This daytime episode transformed a dream into a hallucination, which made them both worry. Donal's wife urged him to seek help.

"What do you think is causing it?" I asked.

"I had an aunt who started seeing things and she was doolally within six months. Alzheimer's," his wife said.

"Certainly diseases like Alzheimer's can cause people to hallucinate," I answered, "but not usually in the early stages when your memory and faculties seem as sharp as ever." I had tested Donal's memory and decision-making and cognitive abilities and they were strong. I turned to Donal, "Is that what you are worried about? Alzheimer's?"

"You're the doctor."

I was the doctor – but I knew I was having difficulty getting past the fact that Donal's first hallucination had immediately followed the discovery that his job was under threat. The practice of medicine requires the doctor to know the facts and figures of diseases and their treatments. The art of medicine follows the narrative and gives meaning and context to the patient's story. I am drawn to my patients' backstories.

"I'm not saying anything definitive now but I think it is worth considering the possibility that these symptoms are related to the stresses at work."

Donal didn't reply.

"It's quite an unusual symptom … and stress can cause some very odd things," I added. I thought I might have seen his wife give a tiny head movement that I took to be agreement. But I wasn't sure so I didn't press on. "Well let's not worry about that possibility now. There are tests that need to be done. Let's see what they show."

I perceived Donal to be guarded. It made sense to me that his visions were a trickle of something desperate to escape. But I reserved final judgment. I arranged some investigations.

Donal had an MRI scan of the brain. It was normal. Seeing imaginary cartoon characters is nobody's definition of normal so the clear scan felt like a lie. Donal then had an EEG. The EEG was normal too.

Spike discharges indicating epilepsy may only be present on an EEG once or twice a day. They are easy to miss. Sometimes they are only visible at the exact moment that the seizure occurs. When the brain is tired or under strain, abnormalities are more likely to be present. I arranged for Donal to have a further EEG. This one would take place after he had been deprived of sleep. He was given instructions to stay awake until four in the morning and then present to the hospital at nine o'clock for the test.

Sleep deprivation is a stressor that increases both the likelihood that the EEG will be abnormal but also the likelihood of a seizure occurring. In fact if you deprive anybody of sleep to a sufficient degree they could have a seizure. Everyone has a seizure threshold, even those without a brain disease. Push the brain far enough with drugs, alcohol withdrawal, lack of sleep, a minor head injury, and a seizure could occur. A person with epilepsy has a low seizure threshold so they do not have to be pushed very far.

Donal's sleep-deprived EEG was normal too.

"Your tests are normal. Good news," I told Donal.

It was only good news, of course, if Donal was feeling better.

"He's had it six or seven more times," his wife told me, "maybe more. I think he keeps it to himself sometimes."

The problem was not going away and I had so far failed to explain it. The brain holds on to its secrets very tightly. Where symptoms are transient there is an added level of difficulty in finding their source. They are already long gone when the doctor meets the patient.

I didn't know what was wrong with Donal. Like many neurological complaints his was unique to him. I cannot imagine that anybody before or since has had his exact experience. I knew my best chance of figuring it out was to bring Donal into hospital and see him when he wasn't well. Donal experienced the hallucination most weeks. I predicted it might be possible for me to witness him having an attack if I observed him in hospital for at least a week. He and I needed only to be patient.

The epilepsy center I work in has a six-bed unit where we admit people like Donal for observation. Individuals are confined to single rooms where they are videoed around the clock. Their brainwaves and heart rates are continuously recorded. Nurses watch them constantly, ready to run into the patient's room if they seem unwell. Most patients stay for five days, some stay for two weeks. However long it takes. The patients never leave their rooms during that time. They can't have a shower or wash their hair. Only the bathroom is private and they are discouraged from spending too much time in there. These people submit willingly, many enthusiastically so, to this intrusion. They have come into hospital with the hope that they will demonstrate to medical professionals what they have been forced to tolerate in private. That way, they will get their longed-for diagnosis, and afterwards a treatment.

I suggested to Donal that I admit him to the monitoring ward. He proved less than enthusiastic at the suggestion, especially when I told him he might have to stay for a full week. He didn't want to take the time off work. Ultimately it was his wife who persuaded him that it was for the greater good. A date was set up and one Monday morning I met him walking onto the ward carrying two neat sports bags, one in each hand. His wife was with him.

"That's not much for a week," I said when I saw him.

He looked nervous. A private man presenting himself to be watched over. The ward staff took him to his room, helped him settle in. Introduced him to his prison. The technicians called to see him and explained again how the test would work. They carefully fixed metal discs on his scalp with glue. In the days that followed they would check the placement of the electrodes regularly and reattach them when they came loose. They were like a large multicolored braid hanging down Donal's back. Each was plugged into a recording device called a headbox which was contained in a small bag that Donal wore around his waist. A ten-foot cable linked the headbox to a computer fixed to the wall. Ten feet would be the limit of any excursion for Donal until the leads were removed again.

To my relief Donal seemed more comfortable once the technicians started setting things up. The equipment interested him. He kept staring at the computer where he could see himself on video with his own brainwave recording running alongside it. The technicians explained the process and asked him to press a button to alert us as soon as he had the slightest sense that the hallucinations were beginning. The instruction seemed to perk him up. It gave him a role in the process. The event-marker button both alerted the nurses that Donal was unwell and also marked a section of the recording for me to review.

"Try not to spend too much time in the bathroom," I told him. "If anything happens in there we won't see it. Did you bring any DVDs or books or anything to keep yourself occupied?"

One of Donal's sports bags contained parts for a scale model of a ship that looked far too big for any table in the room.

"If he takes a liking to you he'll ask you what's your favorite battleship," his wife told me when she saw me staring at it.

"I'll have to think about that," I laughed.

"Is it all set up now?" Donal asked.

"All done. Now we just wait," I told him.

And that is what we did. One day, two days, three days, and nothing. A rotation of nurses sat and watched Donal from a remote viewing station. He knew they were there, but couldn't see them. At first he seemed uncomfortable, aware of being spied on. He glanced at the camera and on one occasion stood under it for fifteen minutes just staring up at it. But eventually, as is usual, Donal seemed to forget about the intrusion and relaxed into the private world that people inhabit when they think they are alone. He talked on the phone, forgetting that we were listening in. "It's not going to happen. I'm wasting NHS money," he said to his wife. The nurses watched him live on camera round the clock. Not a moment missed, because that is our job.

The technicians came in each day and looked through Donal's complete brain and heart recordings. A heart tracing shows the same pattern all day long. It speeds up and slows down but that is the only change. The brain has waxing and waning patterns like a tide. They tell the tale of a person's conscious state. Alert and reading the newspaper after breakfast, they look one way. Drowsy and hungry in the mid-afternoon they are different. Falling asleep they have changed again. Not a second of Donal's brain tide was ignored. Not a single abnormality was seen.

"Sleep-deprive him," I told the nurses when we had waited long enough.

If there were no hallucinations during Donal's stay, his fear that it would have been a week wasted would not be far from the truth. Not exactly wasted from our point of view, because we would have tried our best, but fruitless and frustrating even so. We agreed that he would stay awake until two in the morning and that the nurses would wake him again at six. I hoped that gently tiring Donal would unmask the cause of his hallucinations.

"I'll get a bit more work done," Donal said to the news that he wouldn't be able to while away his time with sleep.

The model ship was taking shape. He was putting small sections together, and if he didn't go home soon, it would become difficult to take with him without damaging it.

"And if it doesn't happen?" he asked.

"We try again. I'll put you on the waiting list to come back in a few months."

We didn't have to. Two o'clock the next day Donal pressed the alarm. The nurses went to check on him as quickly as they could. The whole episode had been recorded by the camera set into the ceiling of his room. I arrived the next morning and found the point on the video when the alarm had been triggered.

Donal was sitting in the chair by his bed. He had the alarm button in his hand and was pressing it. He looked quite well and were he not calling for help I would not have known there was anything wrong with him. I could hear the alarm in the corridor outside. It alerted the nurse who was on duty watching the monitor, and moments later she ran into the room. She was by his bedside in less than thirty seconds.

"They're gone," Donal told her when she arrived. "You missed it."

"Can you give me your full name and address?" she asked him.

He answered correctly.

"And what day and month is it?"

He answered correctly again.

"Why did you press the button? Did something happen?"

"I was asleep and they woke me."

"Who woke you?"

"The things I told Dr. O'Sullivan about."

"The hallucination?"

"Yes. If that's what it is. I opened my eyes and saw something running across the room. They came through the door and ran under the bed. And that was that."

"Are you okay now?"

"They're gone ... *it's* gone. So I'm grand now thanks."

I scrolled backwards through the recording to see what had happened in the minutes before the nurses were called. I pressed play and there was Donal, comfortably snoozing in his chair. Tired from his enforced late night he had fallen asleep with his newspaper on his knee. He looked peaceful. Through the computer's speakers, I could hear the noises of the ward outside his room. I pressed fast forward and only slowed it again when I saw him move. He had woken up. He straightened slightly in his chair and the newspaper slid from his knee onto the floor. Donal didn't move to pick it up. The fingers of his right hand, which were resting on the wooden arm of his chair, began to tap. Then he lifted his hand and the fingers kept moving. It was almost as if he was playing a tiny piano. Then he reached out and picked up the television remote control, also with his right hand. He began pressing the buttons even though the television was turned off. At one point the remote slipped from his grasp but his fingers kept pressing as if it were still there. His head turned to the right and then tracked quickly to the left. That must be Donal following his visitors' progress, I thought to myself. Ten seconds passed before he settled back in the chair again. Every movement he made was quite subtle.

I could see why the person watching the live stream hadn't noticed that there was something wrong. It was only when he had just finished looking from one side of the room to the other that his right hand reached for the nurse alarm and pressed it. Seconds later the nurse was with him. She asked the standard questions to assess his level of awareness. He answered easily.

As soon as I saw Donal's attack for myself I recognized the diagnosis. I had seen many such tapping fidgeting hands, many such quizzical expressions before. To see it for myself was enough, but if it were not, I had Donal's brainwaves to confirm my thoughts.

Every video telemetry patient wears a minimum of twenty-five electrodes on their head. Each electrode represents the part of the brain that lies underneath it. The EEG has its own language to orientate the neurologist anatomically in the brain. The position of the electrodes on the scalp is very precise. Each electrode has a name that is standardized around the world, comprising a number and a letter. An even number indicates an electrode on the right side of the head. An odd is on the left. The letter *Z* is used in place of numbers to denote any electrode placed exactly over the midline. The letters say which lobe of the brain lies approximately under the electrode. The letter *F* indicates an electrode over the frontal region of the brain. *T* is for temporal, *C* for central, *P* for parietal, *O* for occipital and *Fp* for prefrontal (the most anterior aspect of the frontal lobe). Electrode O2, for example, is over the right occipital region, O1 over the left occipital region. When I look at a brainwave recording I use the numbers and letters to know which part of the brain is misbehaving, if any.

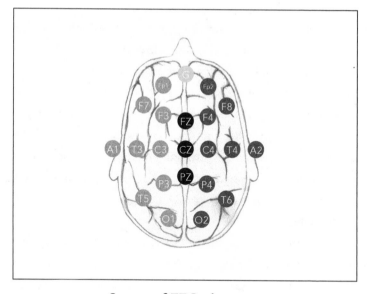

System of EEG placement

I reduced the size of the window showing Donal's video recording. I could still see it in one corner of the screen but, this time, with the brainwaves playing along beside it. That way I could correlate the two. As Donal sat snoozing in his chair the brainwaves showed exactly what they should – a slow wave pattern of normal sleep. When he began to wake, the brainwave pattern changed to one of waking. At first it still looked normal. He was just a man waking from a daytime nap. It was only when the fingers of his right hand had just started playing their imaginary mini-concerto that the proof of his diagnosis showed itself. Out of nowhere the scalp electrodes T4 and F8, over the right temporal lobe, demonstrated a sawtooth pattern. This pattern was not normal.

I glanced from EEG to the video and back again. Donal picked up the television remote control and I saw the unwanted discharge advance as his symptoms did. It moved from neighboring electrode to neighboring electrode on his head, spreading like concentric rings on water. Within a minute it had gone so far that it involved much of the right hemisphere of his brain. Donal was still sitting upright, apparently awake, but not really. Had the nurse known to run and test him at that moment he would undoubtedly have been confused.

Donal's head tracked around the room. The discharge stayed largely confined to the right half of the brain. I waited for it to spread further but it didn't. Instead Donal's expression and posture relaxed. The discharge was only evident for a moment longer. A second before Donal pressed the button to call the nurse, it finally disappeared. It had been there for almost ninety seconds and then the brain had snapped back to normal. When the nurse arrived the seven dwarfs and the unwanted electrical discharge in his brain were both gone.

It amazes me how resilient the brain is, but also how quick one has to be to catch it out. Within two seconds of the seizure being over there was no trace that the wildfire had ever existed. Had I turned my back on the EEG for a minute and a half, I might have been fooled into believing that there had never been anything abnormal to see. Had I tested Donal in the moments just before or after his hallucination there would have been nothing to find. In the thousands of minutes of recording we had made of Donal there had been only ninety seconds in total that proved he had epilepsy.

I went to tell Donal the result. His hallucinations were a symptom of an epileptic seizure. The electrical discharge spread only so far and then disbanded, which is why he didn't collapse or have a convulsion.

"Epilepsy? Seizures? You're sure?" he said when I told him.

"Yes, the test was conclusive."

"But why has it happened?" he asked.

I didn't honestly know. There were no clues. Donal's scan was normal and he had no accidents or illnesses in his past that might explain why this had come about. It would have to remain a mystery. At least until the next set of technological advances comes along.

"I thought epilepsy meant a person had blackouts and that sort of thing ..."

"Hallucinations are known to happen in seizures, although I'll admit this is more specific than any hallucination I've ever heard described before."

I had been confused by the sheer strangeness and specificity of Donal's cartoon visitors. I shouldn't have been. Certainly cartoon dwarfs are likely to be specific to Donal, but there were lots of similar precedents to draw on. Wilder Penfield himself had several patients who reported seeing scenes just as unusual as Donal's. When Penfield stimulated the temporal lobe of one woman she reported a vivid hallucination in which she could see herself giving birth. One man saw himself standing on a street corner. Another man said he heard an orchestra strike up. Penfield tested the validity of these unusual claims by stimulating the patients' brains when they weren't expecting it.

He also occasionally told them he was giving a stimulation when in fact he was not. The visions were stable, only there when he stimulated and only if he stimulated a very specific spot. Penfield could only speculate as to the source of these hallucinations. He guessed that they were memories and likened the brain to a movie camera.

Hallucinations are sensory experiences conjured from nothing. They are actually quite a common symptom in seizures. Most hallucinations are simple – smells, seeing colored dots or lights. The olfactory region where smell is processed is particularly vulnerable to producing seizures when damaged. A frequent seizure aura is the smell of burning, or of rubber. Hallucinatory tastes – metal in particular – are a feature of a seizure discharge in the gustatory cortex responsible for taste perception. The occipital lobe region produces simple visual phenomena such as dots moving across the field of vision. The simplicity of the hallucinations produced by the primary visual cortex is a reflection of the basic visual processing that takes place in that area. The variety of symptoms that can arise from a diseased temporal lobe reflects the breadth of what it does.

I explained all of this to Donal.

"These are seizures. It's unusual but almost any sort of hallucination can occur in a seizure. The EEG recording is quite clear."

"But seven dwarfs ...? Why in the wide world?"

Why indeed. I was as surprised as Donal at the bizarre nature of what he saw. Donal wanted a fuller explanation. I wished I could supply that. He was a person who unpicked things and paid attention to details and tried to understand things completely. Working with brain disease I have become used to never having all the answers. I had to get that across to Donal.

"I can only speculate, I'm afraid. I have seen the abnormal electrical discharge on your EEG so the diagnosis is no longer in doubt. The 'why dwarfs?,' however, is open to possibilities."

*

Most patients who see detailed scenes in seizures have temporal lobe epilepsy. It follows that their cause must be a disturbance of some normal function of the temporal lobe.

The earliest recorded operations to remove a lobe of the brain took place in 1891 in Switzerland. The patients had psychiatric problems and they had parts of the temporal, frontal and parietal lobes resected. This sort of psychosurgery – destroying brain tissue in order to treat a mental condition – was regularly performed from the mid-1930s until the 1950s. In 1949 Egas Moniz, a Portuguese neurologist won a Nobel Prize for perfecting a technique that essentially destroyed the frontal lobes. What is most astonishing about it is that the surgeons, neurologists and psychiatrists carrying out the procedures had no idea what individual lobes of the brain do. This surgery took place before brain function had been properly mapped. They operated blindly. They injected toxic substances into the brain. They drilled holes in the skull and severed connections between lobes.

These procedures were used as a treatment for both epilepsy and psychiatric complaints. (Or perceived psychiatric complaints would be more accurate – willful daughters and disobedient or unfaithful wives were amongst the many victims of psychosurgery.)

A milestone for learning about the temporal lobe came in 1953 when a maverick surgeon removed both temporal lobes of a patient called Henry Molaisan – commonly referred to as patient HM. After the surgery Henry lost all ability to lay down new long-term memories. He could remember details from his past but never learned anything new. Henry became a curiosity for doctors. He lived out his life in institutions where he was extensively studied by neurologists and psychologists. It was through Henry that science first confirmed the importance of temporal lobes to memory, and the hippocampi in particular.

The hippocampi – named after their shape, that of a seahorse – are part of the cortex, which is the surface layer of the brain. The cortex is the organ of thought and intelligence and memory. It can be roughly divided in three. The neocortex is the largest section. Only mammals have a neocortex. It has multiple layers of neurons and it is its complexity that gives us our complexity. The neocortex supplies our higher cognitive functions. Then there is the olfactory cortex, one in each hemisphere. They have only two layers of neurons. They lie on the medial part of the temporal lobe and are involved in processing smell. Finally there are the hippocampi which also lie on the medial surface of each temporal lobe. They have a single layer of neurons folded in on itself. It was the removal of both hippocampi that left HM with severe anterograde amnesia. Memory is stored on both sides of the brain so a person can live with only one hippocampus if it is healthy. The loss of both is devastating.

The pattern of Henry's memory problem revealed new insights into how memory might be stored. He was still able to carry out tasks such as playing games, reading, dressing himself and washing and so on, which meant that procedural or implicit memory was preserved. Thus memory for these sorts of automatic learned tasks and motor skills must lie somewhere other than the hippocampi. Also, he had preserved memories from childhood, so long-term autobiographical memory must also exist outside of the hippocampi. He also had relatively preserved short-term recall or working memory. What Henry's specific deficit revealed is that the hippocampi are vital structures in converting short-term memories into long-term ones.

A great deal of work has gone into the development of our understanding of memory since Henry's unfortunate operation. The brains of people with epilepsy remain a fertile source of information. In 2005 eight patients undergoing explorative operations for the treatment of epilepsy had microelectrodes placed in various locations in their brain. The scientists then showed the patients a selection of random pictures and measured the electrical activity coming from the neurons being monitored. It was possible to demonstrate that specific neurons seemed to respond to specific pictures. For example, in one woman a single neuron fired every time she was shown a picture of the actress Jennifer Aniston. Several different pictures of the actress managed to elicit the same response. That "Jennifer Aniston neuron" did not show activity at other times. This supported a preexisting theory that specific neurons represent the memory of specific objects, people or concepts.

Of course it is probably not as simple as this study makes it sound – one cell for one memory. For a start, some believe that there are actually several copies of each piece of information to stop it getting lost. Also it is almost certainly the case that complex memories are not stored in a single cell or group of cells in a single location in the brain. Most memories are stored in the connections between groups of cells with each group of cells contributing to the richness of an experience. Different aspects of a memory are stored in different locations. For example one area of the hippocampus is credited with allowing us to navigate, while another area of the temporal lobe recalls a piece of music. In remembering something the brain may need to draw on sound, vision and smell and reconstruct the memory piece by piece from information stored in different brain areas. That is one of the reasons that memories are so unreliable. To remember something the brain must replay a pattern of neural connections that originally occurred in response to a particular event. Those connections between cells are unstable and are subject to change every time they are activated. Not every replay is the same – each risks adjusting the memory just a little.

Penfield thought that the temporal lobes contained memories that were recorded in a continuous stream that was faithful to the original. He believed that stimulating the temporal lobe resulted in a release of those memories. It is certainly possible that Donal's cartoon visitors may have been a memory. Perhaps he has a "seven-dwarfs-specific neuron" stored in his long-term memory, from a childhood experience he thought he'd forgotten, or maybe from time spent with his children when they were young.

Or maybe they weren't a memory. Maybe he conjured them from his imagination. Memory and imagination are closely related. I did not need to have seen Donal's cartoon visitors to imagine what they might look like. I had a picture of them in my mind as soon as he told me about them. Functional MRI studies show that imagination is also associated with activation of the hippocampus. It is interesting that several of Penfield's patients saw themselves from the outside in their hallucinatory scenes. This is called autoscopy. Does a woman recalling her own experience of childbirth typically take the perspective of an outsider? This seems unusual for a memory. Perhaps the stimulation of imagination is a better explanation. The temporal lobe is engaged in complex visual memory tasks – storing the memory of familiar objects, faces, shapes and colors. Perhaps the electrical discharge of the seizure created something new for Donal rather than drawing on something already in existence. An electrical surge through the temporal lobe can make the world seem a very strange place.

For the moment we cannot know. Maybe the next generation of technology will give us another more useful puzzle piece. For his part Donal thought long and hard trying to dredge up where he had seen those figures before, but he came up with nothing. Fortunately treating Donal didn't depend on us knowing everything. Donal's seizures went away with medication. He is well so I do not see him anymore. He never did ask me which was my favorite battleship. Probably he had figured out that there are a lot of questions to which I have no answer.

4

MAYA

More miracle than bird or handiwork.

—W. B. Yeats, "Byzantium" (1933)

Fasting, exercise, drinking vinegar, provoking sneezes and the purging of phlegm – each of these things has been advocated as a treatment for epilepsy. Dietary advice, cupping, rest, herbal remedies have also all been recommended. Thankfully we are now in the twenty-first century and things have changed a lot. A large selection of medications are available for epilepsy. These are good enough to allow people to live a normal life. But they don't eradicate seizures, they simply control them. Children can grow out of epilepsy. Epilepsy is often a lifelong condition for most people who have their first seizure in adulthood. However, the good news is that most will stop having seizures as long as they take their medication.

The adult brain has only a limited capacity for recovery, so a neurologist's role is often focused on finding ways to protect and preserve brain cells, rather than on a cure. If a person has a stroke all the treatment strategies available are designed to remove any blockage and return blood supply to the brain as quickly as possible. That prevents damage. If the stroke is in the past there is nothing available in current medical practice to heal whatever has been lost in the brain. Treatment is damage limitation. Management strategies for Parkinson's disease, head injuries, multiple sclerosis and infections are also aimed at protecting the brain. Neurologists are in the business of trying to preserve brain cells because nobody knows how to replace them. We are as bad at curing brain disease as we ever were. Almost.

I have known Maya for over ten years. Her epilepsy long predates our first meeting. In fact Maya's epilepsy is older than I am. Maya was fifty-nine when we first met in an outpatient room in East London. She is a small, delicate lady born in Uganda. She is softly spoken and always carefully turned out, very polite and gentle, but I have learned over time that her demeanor belies her inner strength.

Maya always comes to clinic with her husband, Emmanuel. She is a housewife and he works in a supermarket. The shop he works in is near where I live. He, like Maya, is endlessly pleasant. I look out for him when I am in the shop but our schedules must be different because I have never seen him there.

Maya developed epilepsy when she was ten years old. She could never say exactly how it all started. It had been part of her life for as long as she could remember. She never asked her parents about it when they were alive. When she and I first met there was nobody left to ask. In fact Maya could tell me very little about her epilepsy. That was not her fault. Her only experience of her seizures was waking up from them. Sometimes she found herself lying on the ground. Or she missed half an hour of a television program. Sometimes, alone in the kitchen, she found a spill that couldn't be accounted for: a pool of milk on the table or on the floor. Or she suffered disjointed time – washing up one moment, in the bathroom the next.

Her husband filled in the blanks for her, and for me.

"She goes a bit funny," he told me. "It lasts a few minutes if it's not so bad, or much longer if it's a bad one."

"What exactly does she do in them?" I asked.

"Most of them aren't like normal seizures. She doesn't collapse, only now and again. Mostly she just loses it for a while."

"Does she talk during them? Or do anything with her hands or feet? Or with her face?"

He thought for a while. They had been married for over thirty years. He had seen hundreds of her seizures. That does not make them easy to put into words.

"I'll tell you the best way to explain it," he said. "In the shop there are always people who can't find their wallet. That's what she looks like! They go through their bags in a panic. They try every pocket and they keep pulling things out and putting them back again. You see?" he smiled triumphantly.

I did see. It was a wonderful description. I could picture exactly what he meant.

"Does she talk during them?"

"She's just not there. There's no point trying to talk to her because she wouldn't answer."

Not every seizure was like that, however. Those ones happened once or twice a week at least, but every few months she had a much bigger attack.

"The big ones are the ones that scare me," he said. "She falls over and she stops breathing. Bubbles come out of her mouth. She looks like a dying woman."

"How do those ones start?" I asked.

"Just like the other ones, but they don't stop."

"Can you tell when a small one is going to become a big one?"

"Some of the time. Her eyes sort of widen if it's going to get worse. Waiting to see where it will go is the worst thing. It's a relief when I see it pass and I know it won't be a bad one."

There was no doubt about the diagnosis, my questions were trying to find the root of the seizure in the brain. Maya did not have any odd feelings or warnings. Emmanuel could not say if one limb moved more than the other or if the mouth or face did anything in particular. All we knew for sure was that each attack lasted a matter of minutes. Each was accompanied and followed by profound confusion. Maya always seemed fidgety and anxious and scared, although she never remembered the fear afterwards.

There is no record of what exactly was said or done when Maya was first diagnosed in the 1950s in a small town in Uganda. The first medication to treat epilepsy had been developed decades before. Maya's family were relatively well off in their community but it is still doubtful that treatment would have been available to her where she lived then. Epilepsy was and is more common in less developed countries. Diseases like malaria and cysticercosis infect the brain and the damage they cause leads to seizures. Yet, despite this high prevalence, most of those affected by epilepsy in countries like Uganda will not have access to the appropriate treatment. Even today.

It was when Maya moved to the UK in the 1960s with her family that she first started treatment. She was prescribed pheno-barbitone, one of only a handful of epilepsy medications available at the time. It reduced the number of seizures, although it did not fully eradicate them, and the side effects meant that she felt drowsy and dulled all the time. She had finished school shortly before the move to London and had hoped to go to college, but the seizures and the tablets left her struggling to keep up with the person she had been. In the end she stayed at home to help her mother and watched as her younger siblings passed her by.

For several years after moving to the UK, Maya took escalating doses of phenobarbitone. When that didn't help she took a second drug. She struggled to take the tablets. She barely felt the presence of her own seizures but the unpleasant side effects of taking the tablets she knew all too well. Sometimes she wondered if the seizures were a trick played on her by her family to trap her at home. That sort of thoughts made taking tablets feel like an unnecessary cruelty. She felt well when she didn't take them – so sometimes she just didn't. She got up in the morning, took two tablets from their container and shoved them down the plughole of the sink. Then she turned on the tap and let the water run. When she did that she felt better and nobody else seemed to notice. But then eventually, she would wake on the floor of the bathroom, bruised and with a painful swollen tongue. Or her mother would see her having a seizure and get upset and insist she see the doctor. Maya would find herself forced to admit that she had stopped her medication and would promise never to do it again.

This pattern of stopping and starting treatment recurred for a few years until something happened that frightened Maya into never missing the tablets again. As she recalled it many years later, she had been walking down a familiar terraced street towards a grocery shop she visited a few times a week when suddenly she was standing frightened and alone on a road she had never seen before. Even the style of house was unfamiliar to her. Her shopping bag and purse were gone. A woman was standing in a doorway watching her. Maya had no idea where she was or how she had come to be there. She didn't know what time it was or how far she had traveled from home.

"You okay love?" the woman asked.

Maya understood what was being asked but she couldn't frame the words to answer. Another woman came out of a house to join them.

"Do you know her?" the second woman asked the first.

"You look lost," one of them said, but Maya still couldn't voice her predicament.

"Maybe she doesn't speak English," the second woman offered.

"Yes ..." Maya struggled to get the single word out.

"Yes, you don't speak English, is that it love?" the first woman asked.

Maya just started to cry.

My patients often tell me stories of the unpleasant experiences they have when a seizure happens in a public space. These stories can be so upsetting that sometimes I feel that these are the only sorts of experiences out there. But they're not – patients are usually more likely to be met with kindness than cruelty.

The women helped Maya. They could tell that she was scared and unwell and the immediate reason didn't matter to them. They invited her inside. The Maya I know, who is in her seventies now, is very delicate and girl-like. I can easily see the much younger woman in her and imagine how vulnerable she must have seemed at that age. The women gave her a glass of water and debated what to do. They gesticulated with Maya, finding a language they could communicate in. The women pointed at the telephone. Maya shook her head. Even if she were able to make the call, her family did not have a phone. In the end the problem solved itself. Maya's head began to clear. Time and words began to slip back into place. Five minutes after sitting down she was able to account for herself. She explained that she had epilepsy and must have had a seizure. She discovered that she was not as far away from home as she had feared. When she was well enough the women walked her home. She never found her shopping bag or her purse. Nor did she miss out on taking her medications ever again.

Fifty million people in the world have epilepsy. Nearly three and a half million people in the United States are affected. Of those who receive the correct treatment seventy percent will go into remission. Their seizures will stop although many will have to take tablets all their life. Maya fell into the thirty percent whose seizures continued despite trying many different medications.

Sometimes treating epilepsy is easy. You prescribe a drug and the person gets better. Unfortunately it doesn't always go that way. When the first drug doesn't work, you have to try a second. Then a third. If the third drug is ineffective then you know that you have your work cut out. Long-term follow-up studies observing the progress of people who have to take more than one medication show that every subsequent drug is less and less likely to work, but because seizures are so disruptive to a person's life, many people will tolerate taking a dozen drugs or more. Even when they know the odds are minuscule they keep trying. The drugs can come with horrendous side effects so when they are ineffective that is doubly galling. People respond differently to medications. One drug suits one person but not the next. If a person has very frequent disabling seizures they will often try everything new that becomes available. Every new treatment brings the potential for improvement, but there are also unpredictable dangers. In the 1990s a new drug called vigabatrin came onto the market. Extensive clinical trials had proven its effectiveness. It had been tested to all the standards expected for a new drug and it had the usual slew of minor side effects – drowsiness, dizziness, and so on. It was not until a large population of epilepsy patients started taking the drug that it was discovered that it caused retinal toxicity. The result was significant visual impairment in people already disabled by seizures. This is a rare horror story, but it is still a reminder that for some people it is wise to stop changing treatment if they are coping well.

Maya had tried six different tablets. One day she realized that the tedium of the drug changes was taking up more of her life than the seizures. After that she refused to try anything new. The medication she was on at that time had reduced the attacks by half and she resigned herself to the fact that she would never get any better than that.

By the time Maya and I met she had had epilepsy for almost fifty years. She lived with it and tolerated it. She seemed to me to be a woman who had had a good life – not perhaps the life she would have had if she had not developed epilepsy, but good nonetheless. She didn't have the opportunity to further her education and was never well enough to get a job, but she had replaced these losses with other things. She had found plenty of happiness in her choices. She married and had five children of whom she was very proud. The success of the family fulfilled her. Once she had decided that having epilepsy was her lot in life, she had stopped discussing it with her doctor and just got on with living.

Maya had made an appointment to come to my clinic because of two changes in her life. Her children had grown up and left home. She was spending more and more time alone. Maya's family worried about her safety. Then her long-standing GP retired. Her new doctor insisted she have her epilepsy treatment reviewed. Maya walked through the door of my clinic reluctantly and with low expectations. My own expectations matched hers more than she realized. I read through her notes and had seen the limited impact of her treatment. Before our conversation had even started I had it in mind to do my best to avoid messing with a life that did not seem particularly broken.

"Nothing's ever helped her," her husband told me after we had talked about the situation.

"I see that, but there are new treatments now. There will be no miracles but I can try one of the newer drugs if that's what you want."

I was looking at the poised, politely smiling woman in front of me. In my mind I was concerned about what to offer her. A new treatment that could help, or be inert, or take her delicate equilibrium and rock it? I wondered to myself if this woman really wanted to take that risk.

"I've tried everything," Maya told me.

"Believe it or not, you haven't. There are lots more drugs I can offer if you want to try them."

"Are they better drugs?" her husband looked at me with hope that I knew I would immediately dash.

"Not better necessarily, just different. They certainly have fewer side effects than the older drugs so they are generally nicer to take."

"They are nicer to take?"

I was offering hope but with contained expectations. Any drug that has an effect on the brain is bound to come with the potential for unpleasant side effects such as drowsiness, loss of balance, memory impairment, personality change, aggression, anxiety or depression. When you manipulate the brain you risk altering the fundamentals of a person.

"What do you think?" Maya's husband asked her.

"We'll do what you say," Maya said to me.

"You know how badly these seizures are affecting your life. Are there a lot of things you want to do but can't?" I asked.

The couple looked at each other. Their faces said that of course there were things she couldn't do.

"You came here today so I assume you wanted to know what else was possible," I said when I hadn't had an answer.

"A new drug might work or might not? Or make me worse?"

"If it doesn't work we can always stop it again. Of course I would never make you keep taking anything you're not happy with. But starting and stopping drugs takes time."

"Tell me Doctor," Maya said, "there is still no cure? After all these years?"

She put the question as if it were an imposition to even ask. I felt a pang of guilt. I hadn't told her about all the options. It was our first meeting. I had no real sense of how bad things were for her. Nor did I know her goals nor how brave she was willing to be to get there. She seemed to me to be a woman who had lived with an unstable brain disease for most of her life and who might just very well be better off continuing doing just that. But maybe this latest question suggested that bearing her burden was not as easy as she made it look.

"Well ... there is another option. It is only suitable for a small number of people and it's quite a big thing: surgery."

"Surgery? On the brain?!" her husband said.

"Yes."

"Oh no Doctor, we don't want that," he said, "no, no, no."

Surgery is a possible cure for epilepsy in a small percentage of sufferers. It can feel like a miracle when it works, or it can feel like a desperate last-ditch leap of faith that is traumatic and requires sacrifice and still doesn't work. Maya was fifty-nine, old by the standard of those who underwent this type of surgery at that time.

"It *is* a very big deal," I agreed, "but when it works ..."

"No, no, no," her husband kept shaking his head.

"Maya?" I said.

"You said *when* it works," her husband interjected.

"Yes, it doesn't always work, it comes with risks, but if you are interested at all then just looking into the possibility is risk-free. Maya?"

"*I* could have surgery and be cured?" Maya spoke before her husband could object again.

"I'm not really sure at the moment. Maybe. I could arrange for some tests to see if you are suitable. After the tests I can tell you if it is a possibility and how high the risks of surgery would be for you specifically."

I almost hoped she would say no. What if I took a woman who was living a full life and made things worse for her by offering her a brain operation that she could live without? She didn't have a life-threatening brain tumor that had to be removed. Surgery was not essential, it was optional.

"Not everybody is suitable," I repeated, "and it is a *major* operation. It's brain surgery, after all. But yes, to answer your question, for some people surgery is a cure."

"What would it involve?" Maya asked.

That was my first sighting of the strong, resolute Maya, the woman who raised five successful children despite never having had a seizure-free week in their lifetimes. Maya knew something that her husband and I did not – what it felt like to live with the ever-present threat of epilepsy.

"You've had it so long," her husband said.

He had never known her without seizures. Maybe it was because this version of Maya was the only one he had known that he did not see the risks of surgery as worth taking.

"There's no harm in doing the tests and seeing where we stand," I said. "Checking your suitability for surgery doesn't mean you have to go ahead with it. You can change your mind at any time." I looked at the conflicted faces in front of me. "I'd have the tests if I were you."

"Yes Doctor, I'll have them," Maya looked very pleased.

"And maybe we could also try a new medication in the meantime. If it works we might never need to talk about surgery again."

Maya had already taken six drugs, so the chance of a seventh making a meaningful difference to her life was slim. But if I was planning to refer her for brain surgery I wanted to be very sure I had tried every safer option first. Of course, the medication did not work. The odds had always been against it. Meanwhile the path to surgery was being paved and Maya wanted to stay on it. Even if her husband did not want that for her.

Maya had had an MRI scan many years before we met. At that time MRI was new to neurologists. The pictures we marveled at then look rudimentary now. That scan had been normal. The new one I arranged for her was not. Technology had at last caught up with Maya's disease.

Ironically, in the investigation of chronic treatment-resistant epilepsy, we occasionally find ourselves celebrating abnormal tests.

"Good news," I told her, "the scan shows a small scar in the left temporal lobe."

"Scar" is the euphemism neurologists sometimes use to refer to abnormalities on scans. It is an inert sort of word. In Maya's case the left hippocampus looked small and shrunken. It also shone out bright when compared with its partner on the right.

The scan appearance fitted with a problem called mesial temporal sclerosis. (Mesial is a combination of the words *medial* and *basal*, indicating a location on the inner undersurface of the temporal lobe.) Mesial temporal sclerosis is a loss of neurons in the deepest part of the temporal lobe and is a well-recognized cause of epilepsy.

"So you can operate? Take out the scar?" Maya said tentatively.

"I'm afraid it's not that simple."

Brain science is laden with misleading and unreliable information. Practicing neurology does not always feel very scientific. Neurologists have to draw inferences from vague descriptions that defy words and from shadows on scans. It is tempting to think that if a person has epilepsy and they also have a brain-scan abnormality, then the second must be responsible for the first. No such assumption can be made. Amongst other causes, mesial temporal sclerosis may be the result of repeated or prolonged seizures. The scarring might have caused Maya's seizures or her seizures might have caused the scarring; chicken and egg. Maya's epilepsy could be coming from anywhere in her brain, with mesial temporal sclerosis only a by-product. If that were so, then removing it would not make her any better.

"I think this scar might be a cause of your epilepsy, but I need more proof before we can even think of going ahead," I told her.

I hoped seeing her seizures in the video telemetry unit would help to refine her diagnosis. I could use the clinical and EEG clues to try to track the seizures into her brain and hope that they led me to the left hippocampus.

The next time Maya and I met, she was in the video telemetry unit for monitoring

"Thank you, Doctor," Maya said for the hundredth time.

I longed for her to stop thanking me. Doctors are taught to advise patients of all the potential things that can go wrong. Recriminations in the face of dashed hopes can be the hardest to take, so you learn not to be too hopeful. You remember the failures more vividly than the triumphs. Doctors are strategically pessimistic by training.

"What do your children think of this surgery idea?" I asked Maya.

I feared her husband would again try to dissuade her from surgery. He was scared for her. I hoped the rest of the family would support them both.

"I haven't told them yet," she answered.

"Get all the facts first."

"Yes," she smiled.

I was watching a skilled negotiator at work. She sat by her bed with a book on her knee and a stack of magazines on her table. A woman prepared to wait for as long as it took for me to witness her seizures. The windowsill beside her was covered in "get well" cards.

"They know you're in hospital?" I asked her.

"I've told them you're doing some tests."

"No need to be too explicit with them," I said.

"They have their own lives to live," Maya said and gave a small smile.

We settled back into hopeful waiting. Seizures, like buses, like the weather, cannot always be relied upon. But Maya's double-edged luck was still in. In a five-day stay in hospital she had two seizures. She clasped her hands together when I told her. She had suspected that something had happened but had not been sure.

"The nurses kept asking me if I was all right so I hoped for hope's sake that I'd had one. Did you get the information you wanted?"

I did. I had spent the morning looking at Maya's recording.

The first seizure occurred on the evening of her second day. The video showed Maya awake, sitting in her chair looking at the window, seemingly lost in thought, her brainwaves normal. Electrical brainstorms rarely come with a preceding squall. They are with you in seconds. Maya's discharge took immediate hold and its first effect was clear: Fear. Her whole expression was engulfed by it. She didn't reach for the alarm. Whatever she was experiencing had removed her from the world around her.

She began to scream. It was a pleading noise that I had heard a hundred times before. The sound of a seizure can be as distinctive as the sight of it. A colleague and I sit back to back in an office looking at videos of seizures. I can hear what she is looking at but I can't see it. Sometimes when she is watching a seizure I know without looking what sort it is. I have learned to recognize the different war cries. Maya's was the cry of a temporal lobe seizure. I am lucky that I have never witnessed a person in a situation of real terror, but if I did this is the sound I would expect to hear. Fear distilled.

The noise did for Maya what she could not consciously do for herself: it alerted the nurses and two of them came running.

"You're all right, you're all right," one nurse said as she moved Maya's tray-table a safe distance away.

"Remember the color blue," the other nurse said loudly and clearly as she hunkered down by Maya's chair. Maya clutched her, grabbing tightly onto her arm. Her eyes widened, the noise had stopped but she kept looking around her as if something terrible was hiding nearby and she was trying to locate it.

"You're okay, you're okay," the nurse kept saying, "tell me your name, tell me where you are."

Maya didn't answer.

"There's nothing to be frightened of, you're safe here with us. You're fine, you're fine, you're fine," the nurse kept patting Maya on the hand reassuringly.

"Maya, can you tell me where you are?" the second nurse asked.

Both nurses knew that when I reviewed the video some hours later I would need good evidence of Maya's level of awareness. Did Maya know she was in hospital? Could she interact? One nurse took a pen from her pocket and showed it to Maya.

"What is this?"

Maya ignored the pen as if it was not even there. She had stopped screaming, but had started chewing furiously. She reached over to pull the tray-table back towards her and began rummaging through the objects that lay on it. With her left hand she picked up a fork and put it down, and then she took a knife and began to tap it gently against a plate. The nurse eased the knife from between her fingers and moved it out of reach. Maya's searching hand began rearranging a pile of magazines, taking one from the bottom and moving it to the top.

Once that was done she did it again. While her left hand played with everything in reach, her right hand was relatively still and had begun to stiffen. The fingers were very straight and the thumb was pressed up against them, like somebody making shadow puppets on the wall. A duck bill. At the same time Maya's head turned slowly and forcefully towards the right.

"It's going to generalize," I thought to myself. A scream of terror, chewing, fidgeting, hand posturing, head turning. Each stage meant something. I was watching the electrical discharge progress through Maya's cortex. Just before a seizure discharge spreads from being focal to engulfing the whole brain, something very distinctive often happens to the person affected: their head turns purposefully and forcefully to one side. When I see this unnatural head turning suddenly appear halfway through a focal seizure, I prepare myself for the person to collapse and convulse.

Seeing Maya's head turning in the video made me tense up, even though I was watching a day after it had actually taken place. I held my breath, even though I knew she was safe, and felt relieved when seconds later the seizure just stopped. The electrical discharge abated and the fear drained from Maya's face. In a matter of seconds her whole body relaxed and she turned to look at the nurses and smiled benignly at them.

"Maya, are you better? What's this I'm holding in my hand?" said one nurse, showing her the pen.

Maya reached out and tried to take it. The nurse held firmly and asked again, "Do you know what it is, Maya? What's it used for?"

Maya reached up and rubbed her nose with her left hand. She lost interest in the pen and looked around. She let out a small laugh. The nurses asked and asked again: What's your name? What's this I'm holding in my hand? A minute passed before the light of recognition began to show in Maya's eyes.

"Do you know where you are?"

Maya nodded, laughed uncomfortably, but didn't answer.

"What's this called?" the nurse indicated the pen once again.

Maya didn't answer but was paying more attention now. She seemed engaged with the nurse at last.

"Do you know what this is, Maya?"

"Yes," Maya spoke her first word, "It's a … it's a …" she laughed uncomfortably, "I know!"

"Maybe tell me what's it's used for if you can't remember the name?" the nurse tried next.

Maya pressed her thumb and forefinger together and mimed writing and laughed again. She looked embarrassed.

"Good," the nurse said. "Can you point to the window with your right hand and then to the ceiling with your left?"

Maya did as she was instructed.

"What's in here?" the nurse picked up Maya's glasses case and showed it to her.

"Eyes," Maya flapped her hands as if the correct word was just on the tip of her tongue. She took out the glasses and put them on. "Eyes?" she said.

"Can you say the name of them?"

A visibly frustrated Maya shrugged and shook her head and mouthed words she couldn't quite make real. For two minutes things continued like this, until she suddenly said "Write, write," in response to yet another question about the pen. Then a big sigh and she said the word "pen," to everyone's obvious relief.

"Pen" was followed by "glasses" and a minute later she was conversing normally. She told the nurses she was tired and they helped her into bed and very shortly after that she was asleep.

I looked at the video of the second seizure. It was almost identical to the first, except for being a little shorter and without the added feature of the slow, threatening turn of the head at the end.

I looked at the EEG trace. The sawtooth pattern of a seizure. Clear and discrete, it began at the moment Maya's expression first changed. For the first few seconds it remained confined to two lines of a twenty-four-line trace. F7, an electrode that is stuck just behind the hairline and above the level of the left ear. T3, a few centimeters back from that. These correspond with the very anterior part of the temporal lobe. They are as good a representation as it gets of what is happening in the mesial temporal region which is tucked underneath it.

*

Maya's seizure had told me a story in actions. Fear, loss of awareness, fidgeting, arm posturing, loss of speech. Subjective, abstract experiences like fear and other emotions will always be the hardest for neuroscientists to investigate. How do you study genuine terror? Or artificially produce real joy? Functional MRI studies often involve comparing brain responses in large groups of people. But how do you compare people with their own unique emotional life?

What's more, emotional processing occurs in both sides of the brain. If a function is controlled by both hemispheres then damage to one side can be compensated for by the other. Before fMRI it was only if corresponding areas of the brain in each hemisphere were both destroyed that the effect of their loss, and therefore their purpose, could be identified. But the majority of strokes and head injuries that allow people to live on and be examined are unilateral. An accident so severe as to produce damage to both sides of the brain will either be fatal or will produce such widespread disabilities that one cannot be separated from another.

In the case of fear, epilepsy proffered some insight. Fear is well recognized as a frequent symptom of temporal lobe seizures. This created a link between emotional processing and the temporal lobe. Electrical stimulation of the mesial temporal structures – the hippocampi, amygdala and parahippocampal gyrus – can elicit a fear response.

Animal experiments honed this further. The amygdala is an almond-shaped structure adjacent to the hippocampus. We have two of them. Historical studies on macaque monkeys have shown that removing both amygdalae will decrease their attention to threat – they become less wary and are more likely to explore new objects rather than shy away from them. But can you really extrapolate from animal behavior to human? How can you tell if an animal's change in behavior is truly lack of fear? Maybe the monkeys had just lost the ability to recognize objects. Does removing the amygdalae cause a visual recognition problem or an emotional problem?

The amygdalae are difficult to study in humans. Losing the function in both amygdalae is exceptionally rare. But it has happened. In 1994 a woman, who came to be known as SM, suffered selective damage to both amygdalae as a result of a rare inherited disorder. It appeared to make her fearless. Exposed to things that others would find frightening, she instead showed curiosity. Living without amagdylae has not served SM well, however. Fear protects us, and in its absence she has become the victim of an above average number of muggings and other acts of violence.

More recently fMRI has added to what we know about the amygdalae. They are indeed important to fear processing, but it is increasingly clear that the brain cannot be divided neatly into specific centers for specific functions – especially those as complex as emotions. The amygdalae are involved in fear conditioning and the expression of fear but, as with all brain functions, they act in concert with many other brain regions. They are our instant warning system. Sensory signals arrive in them before they reach the more reasonable parts of the brain. The amygdalae react instinctually, after which we rely on the frontal lobes to ensure that the response is socially appropriate. Fear is not the amygdalae's only response. They might generate aggression as easily as making us turn and run (the well-known "fight or flight" response). Fear, anxiety, depression and aggression have all been linked to the amygdalae.

The emotional brain is still being teased out but what is known was enough to suggest that Maya's intense emotional outburst at the onset of her seizure had been a cry for help coming directly from the almond-shaped tissue tucked inside her temporal lobe. But either hemisphere could be culpable. Hard evidence was needed to further implicate the left side.

As Maya's seizure progressed, two very important things happened: she developed arm posturing and loss of speech. When a neurologist approaches a puzzle they distinguish between hard and soft signs. Speech and movement are easy to examine. Also, the cortical area responsible for each lies in only one hemisphere. These features make them hard signs. A level of trust can be placed in their presence or absence.

Historically, loss of language was the feature that first helped confirm the long-standing suspicion that brain functions are localized. Language is a vulnerable function in the brain. It is under the control of the dominant hemisphere – usually the left, if a person is right-handed. It is fairly discretely placed in the brain and has little or no backup on the opposite side, which makes it easy to lose. Deficits in language are obvious and quantifiable.

The ancient Greeks thought language deficits were due to diseases of the tongue. Treatment was directed at the throat and mouth: tongue massage, gargling, and so on. Confirmation that they were unequivocally a function of the brain came in 1861 with two landmark cases. First, in a failed suicide attempt, a man shot away his frontal skull bone. The man's care fell to a doctor who had an interest in the field of cerebral localization. He applied a spatula to the dying man's exposed brain, and found that when he pressed on the anterior aspect of the left frontal lobe he could stop the man's speech. The same did not apply to the right frontal lobe. In the same year another man suffered a stroke and lost the ability to speak. His intellect was intact, as was his ability to understand language. He could gesticulate in response to questions, but the only word he could say was "tan," which became the name by which he was known. Tan came to the attention of a physician and anatomist called Dr. Pierre Paul Broca. Broca studied Tan's speech in life and then, after Tan died, he carried out his autopsy. He found that a stroke had damaged an area of Tan's left frontal lobe.

These two doctors gave the proof that an area of the left frontal lobe is vital to speech production. Later, through similar case studies, an area in the left temporal lobe was proven to be equally important to understanding speech. Functional MRI expanded the subject further. It showed that speech is processed by several connected areas of the brain each representing a different aspect of language production and understanding: naming, combining words, grammar and so on. But it still stands that the position of these areas in only one side of the brain means that neurologists continue to rely heavily on speech when assessing the integrity of the dominant hemisphere of the brain.

Loss of fluent speech, but with retained understanding, is called an expressive dysphasia. It places a disease in the frontal lobe of the left hemisphere of the brain (or it could be the right hemisphere in a left-handed person). Maya had developed a marked problem with speech at the end of her seizure. She could mime the use of objects but couldn't name them. If I wanted proof that Maya's seizure was disrupting her left hemisphere, then this would qualify.

Voluntary movement is another of the neurologist's hard signs. You can see movement. You can test it and reproduce it. The way Maya moved during her seizures was another clinical signpost. Maya's arms had behaved like they belonged to two different people.

Maya's right hand had become rigid and immobile. Penfield's cortical homunculus placed motor control in a strip of cortex in the frontal lobe. The left frontal lobe controls the right side, and vice versa. The stiffness of the muscles in Maya's right hand indicated that they had been electrically activated. Like her loss of language, this placed the disruption in her left frontal lobe.

Meanwhile her left hand picked mindlessly through a pile of magazines. This is referred to as an automatism – an unconscious behavior that occurs during focal seizures. Automatisms may be fumbling movements, plucking at buttons, clicking fingers, tapping a hand. They commonly involve the mouth where they manifest as chewing and lip smacking. Or they may be wild flailing movements – cycling of the legs or frenetic thrashing. More often than not they are seen on the same side as the seizure discharge in the brain. They are purposeless movements where the exact mechanism is uncertain. It is possible they are release phenomena happening as a result of a loss of inhibition of movement in the side of the brain not affected by the seizure.

Maya's puzzle pieces were fitting together to create a picture that made sense. Her loss of speech, her stiff right hand and her fretting left hand were all pointing towards dysfunction in her left frontal lobe. But her seizure hadn't started with those features – it had ended with them. It had started with fear. It seemed the electrical discharge had begun in the region of the left amygdala and then spread forward to involve the motor strip and language areas of the left frontal lobe. This fitted with the theory that her seizures came from the scarred mesial temporal structures seen on her scan. Her EEG change in the F7 and T3 electrodes sat neatly over the left temporal lobe too. The signs were in Maya's favor. It was hard to escape the likelihood that the shrunken left hippocampus was indeed the source of her problems.

I told Maya the results. I could tell that she would have the operation then and there if I offered it. I did not. Everything you do to the brain puts it at risk. The risk is not just physical disability. In the process of removing what is abnormal, brain surgery can disrupt something that technology can't measure or predict. Functional maps of the brain are incomplete. While a surgeon can usually tactically avoid the motor cortex or Broca's language area, they will inevitably remove or destroy some normal tissue. There is no way to know what the effect of that will be. The surgeon could achieve their surgical goal – cure – while inadvertently and unavoidably changing some fundamental part of the person. In one man, a musician, surgery cured his seizures but also affected his ability to appreciate music. His life was irrevocably changed, although nobody looking at him or talking to him could tell how disabled the surgery had left him feeling. A change in mathematical skills or language fluency or temperament – each of these would matter a different degree to each of us. Part of the surgical assessment is to try to figure out what matters to the individual. What can a person tolerate losing?

What if Maya's left temporal lobe was both the source of her seizures, but also sustained her memory? If the mesial temporal region was removed, could her memory disappear along with her seizures? She might exchange one disability for another.

That is very much what happened to patient HM. He had both temporal lobes removed and became amnesic. If a surgeon took out Maya's left temporal lobe and the right was not working well enough to compensate, she could end up amnesic too. Maya was resilient and intelligent and caring. What if some part of that changed?

"More tests," I told her. "Don't worry, we're nearly done."

I referred Maya to the psychologist. This is another way to test the brain. MRI scans and EEGs tell us nothing at all about how well the brain is functioning. They cannot determine how organized you are, how even-tempered, how good at planning, how creative. For the practical aspect of testing our abilities, face-to-face clinical assessment is all there is. If I want to know about muscle strength then I test my strength against that of the patient. If I want to know whether a person can read, I ask them to read. If I want to know if a person can plot a route, I give them a route to plot. There is not a single scan or piece of technology that will answer these questions. They must be tested in the old-fashioned way – by a person.

Determining cognitive function is the job of neuropsychologists. Through a battery of tests in the form of a detailed questionnaire they measure intelligence, memory, language, decision-making ability, planning ability, concentration and attention. I needed the psychologist to pick through Maya's abilities and help me understand what she stood to lose.

After that, Maya had one other person to meet. The psychiatrist. I was on the verge of suggesting that Maya undergo a brain operation. Having a piece of one's brain removed is as major an undertaking as anyone could ever face. Most brain surgery is foisted upon a person as a life-saving procedure. Epilepsy surgery is a choice. Maya had lived with epilepsy for fifty years and could continue to do so. Was she strong enough to make the decision and face the consequences? Even when surgery works it can plunge a person into depression. This may be a factor of the surgical manipulation of the brain. It may be caused by the ordeal of surgery. Depression can even strike people for whom the surgery has been a complete success. If someone has had seizures for most of their life, learning to live without them can be much harder than they anticipate. Or unrealistic expectations might lead to a devastating disappointment.

Once Maya had seen both the psychologist and psychiatrist, we were nearly done. The process had taken a full year. I was happy that it had gone slowly. It gave Maya and her family time to think. There was one final hurdle. Maya didn't need to be present for this one. I was her representative.

Thankfully in high-tech multidisciplinary centers no single doctor can send a letter to a surgeon and say, "Please can you remove a piece of my patient's brain." Decisions like this are too big to be shouldered by one individual. They are pored over in groups. In the hospital I work in, the epilepsy team meets weekly to discuss the surgical candidates. At the meeting there are neurologists, neurophysiologists, radiologists, psychologists, epilepsy nurses, a psychiatrist and a surgeon. Every aspect of the brain has its advocate.

One gray Thursday afternoon I brought all of Maya's test results to that meeting. In a room full of experts we looked at each piece of information in turn. I recounted Maya's story and showed the video of her attacks. I showed her EEG. Then the radiologist projected the brain scan onto the screen.

"Barn door hippocampal sclerosis," he said.

A murmur of approval went around the room. Sometimes we are so used to seeing normal scans that we are pleased to encounter something that can be treated.

"Her verbal memory is already very weak. Visual memory, on the other hand is pretty good. Excellent even. It's her strength. It certainly seems that the problem is localizing to the left," the psychologist summarized her assessment of Maya. We were being told that Maya's left temporal lobe was not functioning well. While both temporal lobes are important for memory, they each have different strengths. The right is for visual memory. The left is for verbal memory. Fifty years of seizures and a shrunken left hippocampus meant that Maya's memory for words was already diminished relative to her memory for what she had seen. In the scheme of things, this was good news. It meant that removing the left hippocampus was predicted to come with only a low risk of significant memory loss. It was already weak and would be compensated by its strong partner on the right.

"She's a very bright lady. High average and above in most tests," the psychologist added.

It didn't surprise me. It also made me feel better that I could defer to Maya's own judgment.

"She feels trapped at home now that her children aren't around so much," the psychologist told us. "She is very motivated to have the op."

"Yes, that's right," I agreed, "she's very keen. She wants more independence."

The decision to have surgery is very dependent on the life of the individual. A person in employment might lose their job if surgery robbed them of some memory or intelligence. They might be more cautious before proceeding. A person in Maya's situation, restricted, could be said to have more to gain than to lose.

"Any psych contraindications?" the chair of the meeting asked.

"I have no concerns," the psychiatrist called out from the back of the room, "she has no psychiatric history and there is good support at home."

"Who will look after her post-op?" the chair asked.

"She has a really great family," I told them. "Her husband is going to work evenings and spend the day with her. Her daughters will take turns to be with her in the evening."

"She's a bit on the older side," a colleague sitting in front of me interjected, "this woman has had epilepsy all her life. Are we really wise to put her at risk of major surgery if she could get on quite fine without it? She's sixty."

"Does that matter?" I asked.

That had worried me from the start. The more seizures a person has the more likely they are to develop memory problems. They could also develop seizures that come from elsewhere in their brain. We try to send most suitable people for surgery as early as possible in their lives, to preserve the brain by protecting it from the consequence of surges of electricity. Also, Maya had lived with epilepsy all her adult life and I feared I was about to compromise the quality of that life.

The chair answered, "We used to have fifty as a cut-off, but not anymore. Outcomes have been just as good for older candidates."

"What do you think?" I looked at the psychologist for reassurance. I was not at all comfortable having my anxiety stirred.

"She has very little memory to lose, is my guess. And the data is in her favor. I think she should have it," said the psychologist.

"I'm just playing devil's advocate," the colleague in front said, and laughed.

"You're not far off sixty, are you?" the chairman called out to him and we all laughed. "Would you have the surgery?"

"Tomorrow!" he said

"Everybody in agreement?" I asked. "Can I give her odds?"

"Seventy percent chance of seizure freedom or significant improvement," the chairman said and heads around the room nodded. Even with all the signs in agreement the science is not perfect. We reserved a thirty percent margin of chance that despite all the tests we were still wrong. The seizures could still be coming from somewhere else in the brain.

I met Maya and told her what had been said. If she had the operation we estimated that she had in the region of a seventy percent chance that her seizures would disappear. She beamed at the news.

"People die in surgery ..." her husband said. Maya's eldest daughter had also joined us this time. She didn't speak, but looked worried.

"It's a major operation of course. There are risks. There is a one percent chance that a life-threatening or disabling complication might happen during the operation." I spoke hesitantly. I knew these risks existed. No patient I had sent for surgery had ever had such a serious complication. Although several have failed to get better or suffered post-operative psychiatric complications, none died or suffered a life-changing disability. But I couldn't help but think that surely every extra person I referred for surgery affected the odds. It would happen eventually. Or did the clock reset each time?

"One percent, that's not bad," Maya said.

"Yes, very serious problems are unlikely but there are less serious complications that are much more common. The memory tests went in your favor but, even so, your memory might get worse."

"My memory is already very poor," Maya told me. "I make lists for everything."

Her husband shook his head nervously.

"Do you want to go ahead?" I asked.

"How do I decide?" Maya asked.

"Well, I suppose if we do nothing things are likely to stay the same. How would that prospect make you feel? Do the risks seem worth it to you?"

"Can you decide for me?"

I did not want to decide. "I suppose ... I can only tell you what I would do if I were you. I would have the surgery. Seventy percent chance of cure is pretty good. But the risks are real so we can't ignore them," I looked at Maya as I spoke. I avoided catching her husband's eye. I sensed, rightly or wrongly, that he would wish that my advice was different.

"How soon can I have it done?"

"I'll start the process. But please keep thinking about it. If you change your mind just let me know."

Maya met with the surgeon for the second time. He discussed the risks again. He described the operation. Three months later Maya was admitted for surgery.

*

Skulls as old as 7,000 years show evidence of rudimentary surgery. This usually took the form of trepanation – drilling holes. Those ancient wounds show signs of healing, suggesting that the subjects of the operations survived in the immediate aftermath of the procedure. There is no way of knowing the purpose of those operations, though.

The first successful resective surgery for epilepsy took place on May 25, 1886, in London. Guided by John Hughlings Jackson, the surgeon Victor Horsley operated on a man who had post-traumatic epilepsy. Horsley removed a visible scar and by all accounts the patient was cured.

Maya's surgery was a refined version of that operation, and the strategies used were very much the same. The difference is that modern neurology works with brain-imaging techniques that allow us to be far more precise than our predecessors could ever be. Modern surgeons can use sophisticated seizure localization techniques to remove pieces of brain even where there is no lesion to the naked eye. But although technology makes the procedure more accurate, it is not perfect, and still works only some of the time.

I remember once reading that we know more about the surface of the moon than we do about the world's oceans, ninety-five percent of which is unmapped. What hides in those oceans?

We don't really know. Maybe they are full of life, or maybe they are not. It makes me think of brain science. Technology creates an illusion that the workings of the brain are more accessible to us every day. They are, but what is not known still far outweighs what is. The brain is full of unplumbed depths. If surgeons can remove sections of brain with no apparent consequences to the patient, one is left wondering what that bit of brain was really for.

A surgeon removed a piece of Maya's brain and we met again six months later.

"No seizures!" Maya crossed her fingers and laughed. She still seemed to be the bright, happy woman I had come to know. Her husband's face carried a soft smile of satisfaction – or relief? Fifty years of seizures, and now they had gone.

Maya had had the anterior third of her left temporal lobe removed. This is a conservative version of the radical surgery that HM and the patients of the twentieth century underwent. It preserves as much of the temporal lobe as possible. Wernicke's area, the brain region that allows us to understand language, is in the posterior part of the left temporal lobe. It is what neurologists call "eloquent brain." Nobody could live a normal life without it. Maya's surgery was carefully tailored to avoid such important areas – or those we know of, at least.

"I'm so thrilled for you," I told her.

"Not too bad?" she said, smiling and patting the hair that was shorter on the left than right. The scar was invisible underneath the regrowth. She was referring to the cosmetic aspect of her recovery but I was marveling at how little the operation seemed to have affected her. Whatever Maya's now missing piece of brain had done there was no evidence of its loss immediately on show.

"How's your memory?" I asked her. "No worse?"

"No problem. I feel good."

I looked at the neuropsychologist's report of the further battery of memory tests done after the operation. They indicated that Maya's memory had not deteriorated and may even have improved. I could only assume that Maya's mesial temporal region had been diseased for so long that it had ceased to perform any useful function for her. Some other bit of brain had taken on its role.

"I'm so pleased for you," I said, but, ever the consummate cautious doctor, added, "Maybe ease yourself gently into the shopping trips, though? Don't you think?"

A shopping trip. The simple pleasure of it. The first time she did the family's weekly shop on her own after the operation, she felt empowered.

Before surgery I had asked Maya what she hoped her life would be like if the seizures stopped. I had asked it casually but the answer mattered to me. I was testing her. I needed to be sure that her post-surgical ambitions were realistic. Life without epilepsy would not be perfect, any more than any life is. Curing her seizures might not transform things in the way she hoped. Her answer had put my mind at rest.

"In my culture, Doctor, when there is a wedding or a big celebration we hire a room or a tent and let people know there will be a party and anybody who wants to come, comes. You don't have to send out special invitations like in the English tradition. Anybody can turn up to our parties. The mother of the family is in charge and the female relatives cook. When my oldest daughter got married her mother-in-law was in charge of cooking. Everybody was worried I'd have a seizure. If my seizures go away I want to do my own shopping, cook the dinners without somebody peeping over my shoulder and I want to have celebration dinners that are my own."

What she said made me feel tearful. It was such a small ask. Now here was the post-surgical Maya looking triumphant.

"I can plan!"

The traditional role she had longed for was finally hers. She packed her small East London house with extended family and friends. She shopped, cooked and hosted, and only asked for help from her family when she wanted it, not because it was foisted upon her.

Predictability has the reputation of being dull. But unpredictability and loss of control are some of the very worst features of epilepsy. I had worried that fear of the unknown would stop Maya even if she no longer had seizures. Having lived with constant threat all her adult life, being released alone into the world proved no challenge to her at all.

Several years have passed since Maya's surgery and she has remained totally free of seizures. She is cured, as far as we can tell. I still meet with her once a year. Last time, she invited me to come to one of her dinners. I longed to go, but she's my patient not my friend so I declined. But I hope instead that someday I'll bump into her outside the hospital.

5

SHARON

Symptoms, then, are nothing more than a cry from suffering organs.

—Jean Martin Charcot (1824–93), neurologist

Sharon was on the Underground when she first collapsed. The circumstances terrified her. It was rush hour. The train was very full and she was standing with several strangers pushed up against her. She knew that she wasn't well. She had several stops to go. She was jammed in at one end. To get off would have required shoving past already disgruntled commuters. And, if she succeeded, she would find herself stranded at a station that was a couple of miles from work. Then, if she started to feel better, she would find it hard to get on the next train. So she took a calculated risk and stayed where she was, hoping to get a seat. Instead she lost consciousness.

"I felt it starting," she told me, "everything went black. I couldn't see anything. I began to panic and I tried to say something to someone but the words wouldn't come out."

Sharon only had secondhand accounts of what happened next. Other passengers realized that there was something wrong when she fell against a man beside her. He grabbed hold of her and the pushing crowd held her upright. It was three minutes between stations and her fellow commuters said that she was completely unconscious for most of that. Once the train pulled into a station the alarm was activated and Sharon was carried to the platform.

"There was a nurse on the platform. She was there when I woke up. She said that I had had a seizure."

Several witnesses said that as Sharon was being carried out of the train carriage her whole body had gone stiff and started shaking. Sharon woke up lying on cold concrete. Somebody's jacket was under her head and strange faces hung in the air over her. The nurse was gripping her wrist, feeling for her pulse.

"They couldn't find my pulse for a while. They thought my heart had stopped," Sharon said.

Sharon was taken to hospital. During her time in the ambulance she woke up fully and was almost completely recovered by the time she arrived at the casualty department. A heart tracing, brain scan and series of blood tests revealed nothing abnormal. Based on witnesses' accounts, Sharon was told again that she had had a seizure, but that it did not require immediate treatment.

Sharon was referred to a neurologist to ensure that advice was correct. She was to expect an appointment in one to two weeks. The day after the collapse Sharon felt very drained. She took that day and then the rest of the week off work. She was due to return the following week, and was beginning to feel better, but on Sunday she had another collapse. This one occurred on a trip to the shops with a friend. Like the first episode, Sharon had felt unwell in the build-up to the collapse. She told her friend she wanted to sit down. They were making their way to a coffee shop when it happened.

"Just before I go it feels like I'm going into a tunnel. Everything goes dark."

In this attack Sharon lost consciousness for several minutes. Her friend told her what had happened. Sharon had dropped heavily to the ground and then lay there, eyes closed, body completely still. Her friend tried to wake her but couldn't. Sharon was rushed to hospital and had a similar array of tests to those administered on the previous visit. Once again she was declared well. Sharon's mother collected her to drive her home. They didn't get there. In the car, Sharon lost consciousness for a third time. Her mother immediately took her back to the casualty department and this time she was admitted.

Sharon stayed in hospital for two weeks. An MRI brain scan picked up a small anomaly: a cyst in her right temporal lobe. The radiologist said it was an incidental finding, not likely to have caused her to collapse. However, finding the cyst did fuel the drive for more tests. A series of specialized blood tests showed nothing amiss. A lumbar puncture was done to look at the spinal fluid for signs of infection or inflammation. It was clear. She had an EEG that showed some irregularities in both temporal lobes. It did not show the spike discharges of epilepsy, but instead some ill-defined changes of debatable significance. Like her MRI scan, it wasn't either clearly abnormal or normal.

The hospital Sharon was admitted to did not have a full-time neurologist. One visited once a week and Sharon had to wait to see him. During that time she had daily seizures. In each she had the characteristic feeling of entering a tunnel and then loss of consciousness before collapsing. The collapses occurred anywhere. Sitting in bed. In the bathroom.

By the time she met with the neurologist Sharon had already had twenty blackouts. She told him all about them and he looked at her test results. Afterwards he said that he thought epilepsy seemed likely, and prescribed an epilepsy drug. She waited on the ward for a few more days. Once the frequency of attacks showed signs of petering out she was allowed home.

The course of the next five years was very erratic and unpredictable for Sharon. At first the drugs appeared to work. She had long periods with no seizures. But eventually the attacks came back and another drug was added. She would get better again, but always only for a while.

"The drugs wear off," she told me. "My body gets used to them and then they stop working."

But epilepsy drugs don't usually "wear off," I thought. Either they work or they don't.

I looked at Sharon's medication list. She was taking a combination of three drugs and had tried three others. No neurologist wants to have their patient on more than one drug, but when seizures are very resistant to treatment it is sometimes necessary. More drugs usually means more side effects. Being on three medications made Sharon feel tired all the time. It also affected her memory. And were she to get pregnant it meant at least a ten percent chance that the baby could have a developmental problem. But Sharon was still having seizures. Concern for her safety was what had shaped her treatment and led her to this position.

Concern was justified. Twice in five years Sharon had been admitted to the intensive care unit. Each time she had had a prolonged seizure that did not respond to the standard emergency drugs for epilepsy. An epileptic seizure that lasts longer than five minutes is deemed status epilepticus. *Status* is a medical emergency. If it is not managed properly it can result in brain damage or death.

When Sharon arrived in the hospital in the grip of these long-lasting seizures she was placed in an induced coma. Her muscles were paralyzed, she was put on a ventilator that took over her breathing, and was given large doses of emergency epilepsy drugs until things came under control.

It was because all efforts were failing that Sharon had been referred to my clinic for a specialist epilepsy opinion. She wanted to find out if any other treatment options were available to her.

"Are the seizures the same now as when they started?" I asked after I had heard the story.

"No. They've changed."

"They've changed twice, I think," her mother said, and Sharon nodded in agreement.

"After a while I started having proper seizures," she said, "or so I've been told. I don't know what's happening in them."

"Proper seizures?"

"She used to just lie still on the ground but about two months in she started shaking in them," her mother said. "Then when she went on the second drug they changed again."

"In what way?"

"She used to have a warning and that stopped. So she doesn't know any more if she will have a seizure. And the shaking became more violent."

"That's when it got really scary. The warning meant I was able to sit down as soon as it started. So I didn't fall. But now I can't make myself safe. I just go," Sharon added.

"The shaking is really strong," her mother said, "she flings her arms out. And her legs kick. If you're standing near her she'll kick you. One time it was so bad it took three of us to restrain her."

"Why do you restrain her?" I asked.

"We have to. If we didn't she'd hurt herself. She once put a hole in the plaster of the kitchen wall."

"Gosh. And how long does all that go on for?" I asked.

Sharon's parents looked at one another and at her.

"Don't ask me," Sharon said.

Her parents weren't sure either.

"Ten minutes," her mother eventually answered.

A ten-minute seizure is exceptionally long. Most are over in a couple of minutes. A couple of minutes can feel a lot longer, especially to loved ones who are looking on powerlessly.

"Are you sure it's that long?" I asked. "Imagine I was counting in seconds during an attack – one ... two ... three – would I get to 600?"

"Easily, I think," her father said, "and some are even longer."

"They're all a bit different. It's so hard to give the right answers," her mother added.

"It's okay. There are no right or wrong answers. I am just trying to get a sense of the attacks. Tell me, do you think her eyes are open or closed during them?"

"Closed, I think," her mother thought for a second. "Yes closed. They roll back into her head before she goes down and then they close over."

"And my memory is getting more and more rubbish. I can have a conversation with Mum and half an hour later I forget I even had it."

"That's right," her mother agreed, "if I ask her to do something she doesn't do it and when I say something about it she denies the conversation ever took place. Once she put a pan on the stove and just forgot about it. I smelled the burning. If I hadn't been at home God only knows what would have happened. I'm afraid to leave her alone."

Sharon's story was a very disturbing one to me. What concerned me most was that I was now convinced that she didn't have epilepsy. Some of the tests were on the borderline of abnormal – but tests must be taken in the context of the patient's story. This story did not say epilepsy to me. I could see no hint that she or her family had any idea that that was something I might say. They had come for a better treatment, not to be told that the diagnosis was wrong.

"What did the doctor who referred you to me say to you about what we might do?" I asked.

I was looking for any evidence that they had their own doubts. My brain was toying with how this difficult conversation could be broached.

"They said she might need surgery, if the drugs aren't working. She has a cyst in her temporal lobe. Do you think that needs to be removed?"

"Cysts rarely cause seizures. They are usually benign, so I'm not convinced that's the problem. There are lots and lots of different sorts of seizures so I think maybe the first step would be for me to see exactly what sort yours are and then plan a way forward. Since you have so many attacks it should be possible to see one quite easily if I admit you to hospital for a few days."

"That would be good," her mother said. "Something really needs to be done. She can't go on like this. She can hardly leave the house anymore."

"Certainly I agree, we need to do everything possible ..." I paused while deciding whether or not to introduce the idea that the diagnosis could have been wrong from the start. "You know, one of the main reasons that people with seizures don't get better on epilepsy drugs is because they don't actually have epilepsy. So that's a possibility. I don't know if it's ever been raised with you before."

"No, it hasn't," Sharon said.

She looked confused.

"Obviously if she didn't have epilepsy we'd be delighted," her mother said, tentatively.

Neither asked what alternative diagnosis I was considering. Surely it was the obvious next question. I gave it a moment but the conversation felt like it had ended. I decided not to take the discussion any further. I had offered them an opportunity and they had declined to take it. This was only our first meeting. Once Sharon was in hospital I would have more time to spend with her. It would be easier to address difficult issues when I knew her better.

I put Sharon on the waiting list and asked my secretary to trace all her notes from her previous hospitals. I was particularly interested in those from the very first presentation to the casualty department. Stories that are told over and over again change with time. Some points are forgotten and others grow.

*

Not long afterwards Sharon was admitted to the video telemetry unit. The junior doctor met with her when she arrived. He explained the purpose of the admission and took some blood tests. We planned for Sharon to stay for five days. It turned out that we didn't need as long as that. Within two hours of arriving Sharon collapsed. It had all been recorded so I could see exactly what had happened.

The technician was called to see Sharon and set up the recording. She measured Sharon's head carefully, marking the exact spots where each electrode needed to be placed. One by one she was fixing the electrodes to the scalp with glue. It was as she was doing this that Sharon had a seizure. The video camera is routinely turned on as soon as the patient is in the room, so it was all recorded.

Sharon was seated in a chair. Her mother, who had brought her to the hospital, was seated on the bed beside her. Sharon was facing the camera. The technician was behind her. When Sharon lost consciousness neither her mother nor the technician noticed immediately. She stayed upright in the chair and the technician could not see her face. Her mother was talking absently to her while looking down at her phone. She only looked up at her daughter when she realized she had not had an answer for a while.

"She's in one!" her mother called out and reached over, placed a hand on her daughter's arm and shook her gently. "Shar? Shar?"

She didn't answer. Her eyes were open and rolled back in her head. The technician stepped around the chair to face Sharon and as she did Sharon began to slide downwards. Her mother caught her by the shoulders. The technician pressed the alarm and then supported Sharon, keeping her in the chair. Sharon's head was hanging heavily to one side. Only the whites of her eyes were visible and her eyelids had begun to flicker. The technician called her name but she didn't show any response.

A nurse ran into the room. She and the technician supported Sharon's body while allowing it to drift gently to the floor. There, Sharon began to shake. Her back arched and her bent legs kicked out repeatedly. The nurse and technician tried to roll her onto her side but the kicking became more violent. They let go and she rolled onto her back again. The staff stepped away slightly and her mother grabbed her legs. She couldn't contain them. The nurse pulled her away.

"Don't. You'll hurt yourself. Just let it happen."

Sharon's right foot hit against the door of the wooden bedside locker with some force. The whole locker shook. Her mother looked ready to grab her again.

"She'll break that door," her mother warned the nurse.

The technician dragged the locker a few inches out of reach. The nurse scattered some pillows on the floor around Sharon.

"I usually give her midazolam when the shaking is as bad as this," her mother said. (Midazolam is a rescue medication that is squirted into the mouth of a person having a prolonged seizure to help stop the attack.)

"It's not a seizure, is it?" I heard the nurse say to the technician.

"No," the technician answered.

Sharon's mother either did not hear or chose to ignore what had been said.

"She'll be fine," the nurse reassured her, "we want to see this so we'll just let it run its course. This is exactly what she's here for."

For the next five minutes Sharon shook and then stopped, shook and stopped again. In between the convulsions she lay with her eyes closed, breathing gently, almost as if asleep. Her mother sat tensely by her side the whole time. She wanted to give the rescue drug. The nurse called the junior doctor who assured her that Sharon's interests were best served by allowing us to video the attack. It took thirty minutes for Sharon to wake and when she did it was with an abrupt startle. Finding herself on the ground Sharon started to cry. The staff helped her back onto the chair and the technician finished applying the electrodes to her head.

I watched the video the next morning.

"I only had half a head on," the technician told me, meaning that she had been only partway through applying the electrodes.

"It's enough," I said, looking at the recording.

I could only see the brainwaves from the right side of the head but they told me plenty. Even when Sharon was unconscious, even when Sharon was convulsing, the brainwaves showed a normal waking pattern.

Her diagnosis was wrong. When a person is having a convulsive epileptic seizure the brain is engulfed in spike discharges. No spike discharges meant this could not be an epileptic seizure. At least one-fifth of people attending the average epilepsy clinic with seizures do not have epilepsy. The commonest alternative diagnosis is dissociative seizures. Dissociative seizures used to be known as non-epileptic attacks or pseudoseizures. Before that they were referred to as hysterical seizures or hysteria. They are seizures that happen for psychological reasons rather than due to a brain disease. Unlike in epileptic seizures, in dissociative seizures the electrical activity of the brain is normal. The mechanism for loss of consciousness is believed to be dissociation.

Dissociation is something that happens to all of us from time to time. It is normal. It is your brain switching off for a moment. You are in a conversation, your mind wanders, you were listening but still you lose the train of what was said. You are reading a page of a book but when you get to the bottom you don't remember a word. In some people it may lead to a feeling of being cut off from their surroundings, or a sense that things seem unreal. Or just a feeling of being spaced out. It can be a protective mechanism – if somebody is in a situation of abuse it can help them to separate themselves from what is happening around them.

Dissociation, when it is pathological, can cause significant illness. That illness can be purely psychiatric or it can manifest as physical symptoms easily mistaken for a brain disease. For some it produces a sense of depersonalization. For others it causes dizziness, blank spells and even blackouts or convulsions. It can lead to poor concentration and memory problems. Dissociation that causes convulsions is as uncontrollable to the affected person as the dissociation that causes us to miss our bus stop or drift away during the news headlines. It is an unconsciously generated process, but it can be overcome.

Everything that Sharon had told me had jarred with the typical clinical features of epilepsy. The seizures lasted too long. They changed forms too often, evolving slowly over time. The convulsions stopped and started. They built up and receded and built up again. An electrical wildfire of epilepsy builds in intensity, spreads and then stops. It doesn't advance and recede and advance again.

When I saw the video of the fall from the chair it confirmed the diagnosis. It had none of the features of a generalized epileptic convulsion, which is usually dominated by the stiffening of every muscle in the body. It often starts with a characteristic loud cry produced as the chest muscles contract and push air out of the lungs. Sharon went limp and fell. Her breath was held and then released with a big gasp. It simply did not look anything like an epileptic seizure.

I decided to record more seizures before discussing the misdiagnosis with Sharon. Because this first attack had occurred before all the electrodes had been applied, and because Sharon had carried an incorrect diagnosis of epilepsy for a long time, I wanted there to be no room for doubt if I were to change that.

"Let's get at least another couple of seizures and withdraw the drugs before I come to a final conclusion," I told the technician.

Over the course of a week I lowered the doses of her epilepsy drugs. I wanted to ascertain if they were affecting any aspect of the test. Drug withdrawal made the seizures worse. In five days Sharon had ten seizures instead of her usual one or two. They also escalated in length and severity. I asked her family to look at the videos of the attacks with me.

"Are these the same as the seizures she usually has at home?" I asked.

"Yes, but she has gone to intensive care when they have been as bad as this in the past."

"How do they compare to the attacks she had at the very beginning? The ones on the first weekend?" I asked.

"They start more or less the same I suppose, but at the end there's more shaking and the shaking is more violent," her mother said.

I had since had a chance to review the notes from Sharon's original diagnosis. At the first presentation to the casualty department Sharon was told by the casualty officer that she had either fainted or had a seizure. He thought a faint more likely but was sending her to a neurologist for their opinion. This original diagnosis had been changed by the sands of time. Sharon didn't recall that conversation.

It was clear that much of the concern about the possibility of epilepsy rested on the witness reports that Sharon had convulsed. Looking through the handwritten notes I could see that the neurologist who met Sharon after she was admitted to the ward had not actually considered epilepsy the most likely diagnosis. *?Pseudoseizures*, he wrote in the notes. This meant that he had suspected that the attacks were dissociative. But then he wrote, *Temporal cyst on MRI. EEG shows some possible irregularities. Treat for epilepsy in the first instance.* He had second-guessed his first impression and decided it was safer to put Sharon on epilepsy treatment than to wait for further tests. Three months later Sharon was feeling better on the epilepsy medication. This was taken as proof that the treatment was working and therefore the diagnosis of epilepsy must be correct. When the seizures came back, and the next medication failed and the next, the initial doubt about the diagnosis was just a scribbled note, long forgotten.

After I had reviewed the video and EEG recording of all Sharon's seizures I arranged to meet with her and her parents. All the seizures had come with a normal waking EEG so I was confident in the diagnosis of dissociative seizures. But I was nervous about having this conversation. Nobody wants to have epilepsy, but a change of diagnosis after five years is a very big thing to face. A change that took her from having a brain disease to a psychological illness – which is still surrounded by stigma – was likely to be particularly upsetting.

I began by explaining that Sharon's brainwaves were normal during all of her convulsions. That was simply incompatible with an epileptic convulsion. It was impossible. When a person is unconscious their brainwaves slow markedly. That is the case irrespective of the cause of the unconsciousness. As a person falls asleep their brainwaves gradually slow and then cycle through different frequencies of slow waves throughout the night. The brain under anaesthetic has a brainwave pattern similar to a version of deep sleep. In a fainting attack the brainwaves slow and flatten due to temporary lack of oxygen to the brain. That Sharon's brainwaves looked like a waking pattern during the time that she was unconscious could only have one explanation – dissociation.

In many ways this was good news. It meant we knew why the drugs weren't working. It offered a fresh way forward. I wasn't sure Sharon would see it that way.

"That can't be right," her father said. "We've been told before, by more than one person, that her EEGs are abnormal. Sharon has a cyst on her brain."

Clinical skills, medical history taking and examination are as important as they ever were in making a neurological diagnosis. Very few people are aware of the dangers of new technology. They think medical investigations can only be a good thing. MRI scans are ordered as a matter of routine now, with only casual reference made to how misleading they can be. MRI scans show the brain in glorious detail never before seen. That includes its imperfections – cysts, atypical blood vessels, unexplained benign spots. People used to live with these without ever knowing they existed. Brain scans are qualitative, and their interpretation will change with the experience of the reporter. It will also be influenced by the clinical information provided by the referrer. A radiologist may dismiss an anomaly if it is nothing serious and couldn't possibly explain the patient's symptoms. For that they need a full medical history for the patient. Often all the information they are given in a referral is a note saying "*?seizure*." Once the scan is reported it is up to the doctor looking after the patient to make sense of it in their clinical context. Much is lost in translation. Sharon's scan showed a cyst. These are benign lesions found on scans that rarely have any clinical relevance. But in a frightened young woman who collapsed on a train, a cyst is hard to forget.

EEG is even more fraught with error and overdiagnosis. Heart tracings are comparable between healthy people. Brainwaves are much less so. Like our external appearance, every healthy person's EEG has the same elements but with great interindividual variation. Not every difference represents an abnormality. Also, brainwaves change with the state of the patient. A drowsy person has slow brainwaves. But the person writing the report based on the EEG findings doesn't meet the patient – what if they don't realize that the patient was drowsy? An "irregularity" is not a diagnostic finding for epilepsy. It is an individual difference. The person who reported Sharon's EEG had never met her or heard her story in detail. The neurologist who received the EEG report didn't see the actual brain tracing.

Doctors are trained to look for disease. They fear missing disease. It comes with recriminations. The original doctor was worried that he might miss a diagnosis of epilepsy. Significantly less fear exists around the neglect of psychological illness. Psychological factors as a cause for physical symptoms are something doctors avoid talking to patients about until every possible disease has been excluded. Until every treatment for disease has been tried and failed. In Sharon's case that took five years and involved six epilepsy medications.

I explained the EEG findings to Sharon and her family. I could see they were processing what was being said. It was unexpected news. They didn't know what to ask.

"But the epilepsy drugs made the seizures better," her mother said.

"I'm afraid that does happen with dissociative seizures. Sharon wants to get better so badly that when she takes the drug it works for a while. It worked because she wanted it to and believed it would. But since she doesn't have epilepsy, the improvement didn't last."

The placebo effect.

"She's been in intensive care twice!" her father said.

People with dissociative seizures are at least twice as likely to be admitted to intensive care as people with epilepsy. There are simple reasons for that. Dissociative seizures last much longer than epileptic seizures. They can last for hours. Most epileptic seizures have stopped long before the affected person gets through the hospital door.

"I think that the doctors in casualty presumed that the attacks were epilepsy and sent you to the intensive care unit for your own safety. They couldn't have known that the diagnosis was wrong. They were doing the right thing with what they knew at the time."

Dissociative seizures in themselves are not dangerous. They destroy lives but don't kill people. A person can have a dissociative seizure that lasts for hours and they won't come to harm. It is the response to them that puts the person affected in harm's way. Once in the hospital, anaesthetised and ventilated, they are at risk of hospital-acquired infections and blood clots. They could have an adverse reaction to the drugs. Sharon was lucky to get out of the intensive care unit without developing a complication as a result of medical intervention.

I had a long discussion with Sharon and her family, during which I tried to allay the assumptions that come with a psychosomatic, hysterical diagnosis. I reminded them that when our hands shake through nerves or our hearts go too fast when we are excited, neither of these phenomena are either deliberate or within our control.

"You are saying there is nothing wrong," her mother asked, "that it's something she's doing?"

"Not at all. The opposite. I am saying that she has seizures, but they are not epileptic ones. Imagine if you had a headache at the end of a stressful day – would you blame yourself for it? Would you say that the pain wasn't real? Physical symptoms that happen for psychological reasons are real."

"For psychological reasons – well what reason then?" Sharon's father was getting angry.

"Honestly, I don't know. At the moment I can tell you that the test is conclusive, but it will take time to understand how this came about."

I didn't know Sharon. Perhaps the seizures happened unconsciously to protect her from something. Sometimes our brains shut down, dissociate, to avoid something unpleasant. In the past seizures such as these were thought to indicate a history of sexual abuse. There is some truth in that for some people – a third of people who experience dissociative seizures have a history of abuse. But two-thirds have not. Other potential causes are loss of a loved one, extreme stress, finding oneself in a trapped situation. Or the seizures might solve a problem in a person's life. Maybe seizures allowed the sufferer to leave work that they hated. Maybe they allowed them to move home to a family who made them feel safe. Maybe they offered protection from loneliness or relationship unhappiness or financial stress.

Or maybe it was none of those. It is a myth to associate all psychosomatic disabilities with life stressors. Sometimes they are a factor of how we respond to illness. They may be part of a cycle of fear and avoidance. I strongly suspected that Sharon's very first collapse on the Underground was a simple faint. I imagine it was very terrifying for her. After it, strangers told her she had had a seizure. Then casualty doctors told her the same. Then she was told that her brain scan was abnormal. Then she was told she had epilepsy. Believing all of that, she must have been anticipating the next collapse. She became frightened and increasingly anxious. She searched herself for symptoms. She worried herself ill.

"I think the first attack you had may have been a faint. The Underground is a common place for that to happen and the feeling you describe of everything going black is very typical of a faint. I wonder if that might not have started it all off?"

"She was told it was a seizure. A nurse on the platform saw her having a convulsion, told the paramedics."

"I understand that but I'm afraid it is very common for people to mistake a faint for a seizure. People shake a great deal more in faints than people realize."

A research study carried out in 1994 proved this point. Experimenters induced healthy people to faint and then videoed the results. When the participants collapsed most had prominent convulsive movements. Faints happen when blood pressure falls. The loss of oxygen to the brain causes loss of consciousness. That causes collapse, which usually results in the person landing prone on the floor. In that position the head is lowered, allowing blood pressure to recover. Blood flow to the brain is restored, and with it oxygen. But where a person is held upright as they faint, the blood pressure takes longer to recover and the faint is prolonged. Circumstances (other passengers) had held Sharon upright when she lost consciousness on the Tube. She had a particularly bad faint as a result, and woke on a cold platform with strangers looming over her. I feared that faint had created a template for future collapses. The problem may have been unwittingly self-perpetuating. She fainted ... it frightened her ... she started to worry she would collapse again ... she did ... she was told she had epilepsy ... her anxiety increased ... and so on.

"Do you have any sense of why they keep happening, Sharon?" I asked.

I wondered about her lack of curiosity when I first suggested the diagnosis might be wrong. Did she have an inkling of a suspicion?

"You're the doctor. If that's what you say is wrong what am I supposed to say about it?"

Sharon brought the conversation to a halt. I discussed the diagnosis further but she and her parents met me with a stony silence. I left them to discuss it amongst themselves. I also asked the epilepsy nurse to talk with Sharon later that day. The nurse represented a more neutral figure than I did. It had the effect I expected. With me Sharon had been stony, with the nurse she expressed the complete range of her anger and distress. She was not upset at the original misdiagnosis. She was furious at the new one that was being foisted upon her.

"She said she'd rather have epilepsy," the nurse told me.

I understood that. Epilepsy is a disease that is relatively easy to understand. It is a disease outside the control of the sufferer. It is treated with a tablet. Dissociative seizures come with no easy solution and the possibility that there was something in Sharon's life that had brought them on.

I referred Sharon to the neuropsychiatrist. She agreed to this reluctantly. Their meeting proved too upsetting and guarded to bear any fruit. The psychiatrist reported that Sharon could not immediately get beyond her shock of the new diagnosis. The psychiatrist would try to see her again in the hope that she would be more receptive with time. I had stopped all of Sharon's epilepsy drugs so she needed to stay on the ward for a further monitoring period.

"She's still struggling," the nurse told me a couple of days later, "and she hates you!"

I spoke to Sharon again. Her mother was with her.

"When she went to casualty one time they said her blood sugar was really low," her mother said. "Maybe she doesn't have epilepsy but shouldn't you be testing her for that? And my husband was talking to friend who is a doctor and they said she should go to see a heart specialist too."

It was difficult for Sharon and her family to face the diagnosis so they were still searching for more palatable explanations. I assured them that Sharon's heart tracing was normal during all of her attacks. The nurses had also checked Sharon's blood sugar and it was normal. But neither mattered because Sharon's brainwaves had already told me an incontrovertible story. A heart problem or low blood sugar would have changed her EEG pattern. There was no change. I explained it all again. She needed to understand the diagnosis was not being made lightly.

"I felt better when I was told I had epilepsy," Sharon told me.

"But you still have seizures, that hasn't changed. Just the name we give them."

"Yes, but at least with epilepsy I knew why."

"You still know why. It's just that the reason is different." .

For five years Sharon had been consistently told that she had epilepsy – so why should she suddenly believe what I had to say?

"I think it's perfectly reasonable for you to doubt the diagnosis," I told her. "Several doctors have told you that you have epilepsy and now someone you have only just met has come along and said that the diagnosis is wrong."

"Why should I believe you?"

"You don't have to, of course, but could we perhaps agree that none of the epilepsy drugs have proved to be of lasting help? You could keep trying new drugs but that hasn't solved the problem so far. For five years you've tried tablets. It hasn't worked. Can I just ask you to consider this new idea for a short period at least?"

"I don't want to see the psychiatrist again."

"If I said you had epilepsy and offered you another drug, would you have taken it?" Sharon didn't answer so I continued. "If you were willing to try lots of drugs even though the first lot hadn't helped, can you consider trying psychological interventions even though we don't know how they will work either?"

Sharon went silent for a long time. Then she gave a shrug that I took to be lukewarm assent.

"Of all the doctors who have given you advice I am the only one who has actually seen your seizures."

"That's true, Shar," her mother said.

It was the first time that anybody in the family had agreed with anything I had said. It felt like progress. She would give the new strategy a few months, Sharon said.

Sharon's situation is not at all uncommon. Traditionally dissociative seizures have been classed as a psychiatric complaint and epilepsy as a neurological one. That has seen people like Sharon discharged from neurology clinics as soon as their diagnosis is discovered. But with no well-established treatment facilities, many of those patients have found themselves abandoned.

Descartes considered the mind and body to be separate. Each could survive without the other. It is rare to find anybody these days who would subscribe to such an idea. It is obvious that the mind and brain are inextricably interwoven, one impacting on the other. Obvious, and yet in practice I think that a lot of people have difficulty sustaining the idea when faced with the practical side of the brain/mind interaction. Organic disease is viewed as more "real," and psychosomatic conditions, such as Sharon's, less "real," less deserving. A person whose legs are paralyzed by a spinal injury is somehow regarded as "more" disabled than a person with psychosomatic leg paralysis. But surely if neither can walk, both are equally disabled. Society does not measure disability by how impaired a person is but rather on a value judgment of the cause of that disability. That is why Sharon reacted against her diagnosis as she did. She knew she would be viewed differently by anybody she told.

Our state of mind is the state of our brain. It is created from biological processes. The mind exists in fluid connections between anatomical brain regions that control memory, perception, imagination, thoughts, emotions, intelligence and beliefs. The mind is intangible and difficult to explain, but it is still real, as are the medical disorders that arise there.

Sharon met with the psychiatrist again and agreed to have cognitive behavioral therapy. A person who has a first seizure is promised to be seen in a first seizure clinic within two weeks. A person with dissociative seizures could wait months for a clinic appointment. Sharon waited three. That was quick. During that time, something very interesting happened: the seizures reduced from three per week to one a month. Nobody had given her any treatment. The improvement was spontaneous.

Several medical studies have shown a similar phenomenon. Explaining a diagnosis of dissociative seizures can be curative for a large number of people. The withdrawal of excessive medical intervention seems to reduce stress levels and takes the focus off the seizures. If you wake up every morning looking for a headache, you might find one. Sharon was getting better on her own.

The problem was not completely solved, however. Sharon had another seizure that just wouldn't stop. Her family called an ambulance. She was taken to the casualty department in her local hospital. With Sharon's agreement I had written to her local casualty consultant. I informed them of the diagnosis and instructed that if she presented with a seizure that she be treated with supportive care rather than drugs. That had proved hard for the casualty doctors in practice. When after one hour the seizure hadn't stopped a junior doctor phoned me and at my request sent me a video of the attack. The attack was the same as those we had recorded in the video telemetry unit and remained entirely consistent with a dissociative seizure. I advised her doctors to continue to watch and wait. An hour later Sharon was sitting up on the hospital trolley asking to go home.

That was the turning point. Waking up in the casualty department and not in intensive care gave her a sense of belief and triumph. When the cognitive behavioral therapy began a few weeks later the progress was substantial. She began to be able to detect minor warning signs if she was about to collapse. The therapist taught her strategies to stop the attack from evolving. They didn't always work, but often they did.

"I can't believe the way it works," she told me when she was finally feeling better disposed towards me. "Sometimes I can't stop the shaking completely, but I can stop myself losing consciousness. I can even restrict the shaking to my legs by doing things with my hands. If I'm sitting at a table people don't necessarily even see it."

Slowly, slowly Sharon was recovering. I didn't ever know the cause with any certainty. I looked for it. The psychiatrist and therapist looked for it. New fragments of Sharon's life did reveal themselves. I began to get a sense that Sharon didn't know how to express her own distress. She had been encouraged from a young age to push through the tough times. I wondered if the collapses had created an opportunity for Sharon to express her feelings and ask for help.

That is conjecture of course. But neurologists are specialists in conjecture. In both epilepsy and dissociative seizures we mostly guess at a cause. It is always interesting to observe how much more easily people accept uncertainty where organic disease is concerned than they ever can when faced with psychological illness.

6

AUGUST

Up from a past that's rooted in pain
I rise.

—Maya Angelou, "Still I Rise" (1978)

"A police officer phoned and asked you to call him back," my secretary told me when I dropped into the office. "It's about August. It sounds like she's been arrested."

"Oh no … poor August."

My heart sank. This was a moment I had anticipated but still I had hoped it would never happen.

"I knew you'd be upset. She's your favorite, isn't she?" my secretary said, kindly.

I started laughing. "I'll go as far as to say she's top ten!"

No parent should have a favorite child and no doctor should have a favorite patient. Still, my secretary's observation was correct. I had a very big soft spot for August. I had known her for a very long time. She was humorous and courageous and I admired her. Not that our early relationship was quite so amicable. We had warmed to one another slowly.

August was and is a clever, rebellious woman, with a mind of her own. Her story begins in a school playground. She was sixteen years old. She was standing at one end of the yard where all the teenagers gathered at lunchtime, her foot resting on a railing, talking with friends. It was an ordinary day. Then something unusual happened. August abruptly stopped talking. She unhooked her foot clumsily from the railing, almost causing herself to fall. Then she ran. She turned her back on her friends and tore across the tarmac. She only stopped when she reached the fence that was the furthest boundary of the yard. Everybody nearby turned to look at her. Not so much because she was running, but because of how she plowed mindlessly through anybody who stood in her way.

The friends August had been talking to assumed that something must have upset her, although nobody knew exactly what. August herself had no sense of why she had suddenly taken off at such speed. Years later I asked her how this episode had felt to her. How had she explained it? She couldn't remember, she said. She just brazened it out was her guess. I think her guess was probably right. The August I know is immensely proud and I can imagine her brushing the whole thing off as casually as she could.

Nothing came of that incident. It was attributed to the behavior of a teenager looking for attention. Unfortunately soon after that August began to take off across the classroom or yard or playing fields in unpredictable but regular bursts. She would stand up suddenly in class and run from the room. Or during a sports lesson she would run around the field with no regard for anybody else.

The problem wasn't seen as a medical one in the first instance. It was assumed that August was acting out in the face of exam pressure. She was an ambitious girl with high expectations of herself. Her mother was informed and August was asked to have a series of meetings with the school guidance counselor. August would not admit to having anything more than an ordinary amount of anxiety about exams. The meetings quickly came to a stalemate and then to an end.

August's tendency to abruptly stand up and run away did not come to an end, however. It got worse. At first her mother blamed the school since that seemed to be the only place where August got this urge to run. But then she started doing it at home too. Watching television, eating breakfast, mid-conversation, August randomly stood and ran from the room. Seeing the attacks in the confines of their own home, August's mother soon realized that this was something more than an ordinary upset teenager. She took her reluctant daughter to see their doctor.

The doctor went through the story with August and her mother. He looked for stresses in August's life and found a few. In particular it was noted that August had not long before witnessed an accident in which her friend was injured. The doctor wondered if this had left her traumatized. Perhaps she was unable to process what she had seen and was acting out instead.

August was referred to a psychiatrist. She concluded that August showed no evidence of any psychiatric complaint. August and her family agreed. Now they were backed into a corner. August had no idea what was happening to her. The school thought the problem was behavioral but had nothing to attribute it to. The GP admitted that it did not seem like a psychiatric problem but wasn't sure what to do next. Meanwhile the problem was getting worse. It was making August feel very exposed, and this made her more and more distressed. She started falling behind in lessons and it became harder and harder to go back to class. Eventually she stopped going to school.

The situation came to a head when August and her mother were walking down their local high street and August abruptly turned and ran out into the road in front of traffic. She narrowly missed being hit by a bus. August's mother and her older brother decided to take her to the casualty department where they insisted that she be investigated.

I can imagine the turmoil of the casualty officer who saw August that day. It cannot have been easy for them to know what to do. Sporadic running for no reason does not feature in textbooks. Some spark of an idea, or perhaps a discussion with a more senior doctor, resulted in a phone call to a neurologist.

As soon as he heard the story the neurologist considered epilepsy a likely diagnosis. He gleaned the fact that August had very little memory of the events. The running episodes were brief. In between them August seemed quite well. All of this fit with self-remitting brainstorms. He arranged a series of tests. Not unusually, they were normal. An MRI brain scan and EEG showed her brain was healthy. Since neurologists expect that tests can be entirely normal even in severe brain disease, he started August on treatment based on a clinical diagnosis of epilepsy. One year and three epilepsy drugs later, August was no better.

By this time it was eighteen months since the running attacks had begun. The lack of response to treatment caused everybody to doubt the diagnosis. August was again referred to a psychiatrist. Then back to the neurologist. The next few years were spent bouncing from neurologist, to psychiatrist, to psychologist, to GP. The diagnosis was reviewed and changed and changed back again. Everybody agreed that August had a serious problem but nothing anybody did made her better. An attention deficit disorder was briefly considered. Her GP knocked that on the head. In the end her attacks were alternatively labeled panic attacks and seizures, but treatment made no difference to August.

I heard August's story for the first time when she was in her twenties. Adele, the specialist epilepsy nurse I worked with, had been called to see her in the casualty department of another hospital. Adele asked if I would admit August to hospital for video telemetry, to try to come to some sort of conclusion about the diagnosis.

The first time I met August she was sitting in the video telemetry monitoring room. As I walked in she was flicking the pages of a magazine aggressively. She looked decidedly unhappy. She had been on so many hospital wards in the previous few years that she had lost all faith that anything positive would be achieved by this new intervention. It was only her trust in Adele that saw her agree to the admission.

I introduced myself and asked her to describe the running attacks to me.

"I just start running around like a nutter," she said.

"Do you know it's going to happen? Do you have any sense of what's happening while you are running?" I asked.

I could see that every new question irked her a little more. I pressed on regardless.

"Can't you read this in my notes? Adele knows all this."

August had correctly identified Adele as someone who could be relied upon.

"Adele has told me all about you, but since we've never met before I wanted to make sure I understood the problem properly."

"I run. What more do you need to know?"

I wondered how many medical professionals she had had to tell this story to. A dozen?

"When the running stops do you know where you are?"

"How many times do I have to tell you that I don't know anything about what happens?"

"I'm sorry. I'm just trying to understand."

"You don't believe me. You think I'm doing it on purpose. That there's nothing wrong with me."

"I promise you, August, I absolutely don't think that. Your other neurologist thinks you have epilepsy and I think he's probably right. I've brought you in here to try to either prove that or get some better answer."

"I've told you everything. Why don't we just start the video so you can see it for yourself. That's why I agreed to come in. Not to be quizzed all over again."

Even with her irritation I wanted to keep asking questions. Diagnosis lives in the story but I had only the smallest sense of August's problem. August's expression told me I might lose her completely if I didn't back off. I decided that was the best course of action. August would stay on the ward for a week. I could get to know her slowly.

"You're right," I said, "let's do the test and see the attacks and then we can talk."

I hoped this would thaw the atmosphere between us. It didn't seem to work. August looked around her room.

"If you think you can keep me in this room when I have a seizure you're mad," she said.

"There are nurses right outside the door, they'll keep an eye on you, so whatever happens isn't a problem."

"This is all such a waste of time."

She had had a lot of tests in the past and they were all normal. Her prediction that this test could be useless too was not without precedent.

"Well it's still worth trying, don't you think?"

August grunted a reluctant agreement. Then she buried her head back in her magazine, giving me no choice but to leave.

The next day Adele called me to let me know that August had had a seizure.

"Can you look at it? They couldn't keep her in the room. I think this might turn into a disaster," she said.

I looked through the video. It was not good. August was sitting by her bed watching television. A tray table sat to her right, between her and an open door leading onto a corridor. There was no preamble. No march of symptoms. When she started moving it was so quick that she was only on camera for a second or two. She stood up and ran to her right. Just ran.

Very fast. That a tray-table blocked her route did not seem to matter to August at all. She pushed through it like it wasn't even there. She was already out of camera view as it rolled on its wheels, leaned precariously to one side and then righted itself at the last second. A plate crashed to the floor. I had only a video of an empty room to watch. The action was taking place somewhere else on the ward.

The headbox she had been wearing around her waist was plugged into a computer on the wall of her room. It had a backup battery so even disconnected from that computer it would continue to record the brainwaves. But in August's case the EEG recording failed. It stopped as soon as she disappeared out the door. The cable must have snapped. I pressed fast forward on the video and watched an empty room flick by on the screen.

The rest of what happened was described to me by the nurse on duty. The nurse's station was a few feet from August's room. The computer that allowed the nurses to watch the video telemetry monitors was situated there. Unfortunately August had moved so quickly and unexpectedly that it took a moment for the nurses to realize what was happening. August was nearly through the double doors of the ward and heading for the staircase when they caught her. They told me later that she was very upset when they found her there. She began to insist that she be discharged. They had to gently persuade her to agree to go back to the monitoring room.

"What's the point?" she said in angry frustration when Adele called to see her shortly after. "If they can't keep up with me and I hurt myself then why am I here? I'm safer at home."

I looked at the EEG tracing of the attack. What was the point, indeed? For the brief seconds before the cable connection snapped the brainwaves were normal. I saw nothing on the video but a woman leaving her room at speed. Adele persuaded her to give us another chance. The technicians replaced the broken cable and we started again.

The nurses were prepared for the second seizure. The attacks typically came in a cluster so we anticipated that there would be another very soon. We moved August's armchair to the side of the bed furthest from the door. That way it would take her longer to leave the room. A student nurse sat in with August. She was there and ready when August jumped to her feet. The nurse was much closer to the door than August so she got there first and shut them both into the room.

Suddenly August had nowhere to go. She swerved and collided with the wall. That didn't stop her. She bounced backwards, reversed direction and headed towards the window opposite. Arriving there she was blocked again and turned to the left and ran the short distance available in that direction. The tray-table had been moved into that corner to avoid another accident. August knocked against it, turned again, ran a few steps and turned a final time. She came to a stop at the tray-table and crouched under it. The nurse followed her but when she tried to get close August shook her arm threateningly, appearing to warn the nurse away. A few seconds passed with August backed into the corner. The tray-table rocked above her. The nurse pressed the alarm and two senior nurses came to join her. By the time they arrived it was over. August had recovered. She stood up, dusted herself off, straightened her clothing, and went back to sit on her bed. It was as if waking up on the floor underneath a table was as normal as could be. The nurses immediately started to question her as part of the assessment of her consciousness and her language. August answered curtly but accurately.

"So you saw it? Now can I go home?" she asked the nurse.

No matter how many videos of seizures I see I suspect I will never stop seeing new things. This one was new to me, unique to August. She was like a silver ball in a pinball machine, bouncing from target to target.

I looked at August's brainwaves. It was hard to read them. August was running so frantically that the insulated leads hanging down her back were bouncing around, pulling on the electrodes on her scalp. This created movement that obscured the recording. Where the brainwaves were visible I couldn't see any unequivocal abnormality. If there was a seizure discharge anywhere in August's brain tracing I couldn't see it. As soon as the running was over and she relaxed the brainwaves were clear and normal again, but by then she was better.

It might be prudent now for me to be more explicitly honest than I have been so far. I may have given the impression that viewing the electrical map of the brain is more reliable and accurate than it really is. To suggest that you can tell everything the brain is doing electrically by sticking metal discs on the scalp is something one would have been wise to be skeptical about from the start. After all skull, hair and muscle lie between the electrodes and the brain. Muscles themselves produce electrical activity that creates distracting noise in the brainwave pattern. If a patient was to tap their tongue rhythmically in their mouth it might register in the brainwaves to look something like a seizure. The EEG electrodes cannot tell if the electrical activity they record comes directly from the brain or somewhere else. It is up to the doctor to differentiate a tapping tongue from a seizure.

What's more, the electrical activity produced by the brain is at a very low amplitude. A whole six square centimeters of cortex must be involved in the synchronous discharge of epilepsy for it to be detected by a scalp recording. Small seizures involving tiny areas of cortex are often undetectable at the scalp. And a discharge buried deep in a brain sulcus is invisible to us. And what of the undersurface of the brain that is hidden completely from view? Looking at an EEG tracing is akin to gazing at the moon on a cloudy night and believing that what you see is enough to tell you about everything that is happening on its surface.

The temporal lobes that sit on the sides of the head are the most recordable. The frontal lobes are very large and have lots of crevices that are impenetrable to scalp EEG. They have a medial and an undersurface that are likewise hidden from view.

A lack of seizure discharge in August's brain did not rule out a seizure. Unlike a generalized seizure, which involves the whole brain so will be detected with only a single EEG lead placed almost anywhere on the head, a focal seizure is only detectable if it is in a location visible to surface EEG.

To make a diagnosis I was dependent on the video and formal case reports of common seizure semiology. In the field of epilepsy there are two categories of brain maps that we draw from. The first attempt to show the functional anatomy of the healthy brain. The second are seizure semiology maps. Seizure semiology is a clinical discipline in which neurologists learn the sign language of seizures.

For the truly unpredictable and multifarious things that happen in seizures there are guidelines. For decades patients have been observed during their attacks and through the correlation of every symptom with the specific brain lesion of each patient we have gained a sense of which seizure symptom arises in which part of the brain. The process is the same as that followed by Hughlings Jackson and Penfield: learning from patients. The only difference now is a bigger pool of patients and the Internet to disseminate the information.

Semiology maps do not include frenetic running amongst them. Like goosebumps and the seven dwarfs, directionless running is too specific to be described in a standard list of epilepsy symptoms. However, there are many case reports detailing other sorts of frantic movements, such as wild flailing, cycling movements and dancing. When they occur, the commonest association is with lesions in the frontal lobes. We refer to these sorts of seizures as "complex motor" or "hyperkinetic."

The frontal lobes are the largest of the brain. They contain many very important functional areas. Amongst them are regions essential to motor control. The cortical homunculus demonstrates the primary motor area identified by Penfield. It corresponds to Brodmann area 4. Neurostimulation of the primary motor area either artificially or in a seizure causes simple movement like stiffening or rhythmic jerking of the corresponding body part. Like the primary visual cortex, the primary motor cortex is involved in the least complicated aspect of movement.

Most movements are not simple. They depend on knowing our position in space, knowing the goal of the movement and which muscles need to contract to achieve that goal. Think of hitting a tennis ball – you need to know where the ball is, where your arm is in space, how to angle the racket, how hard to hit the ball. Hitting a tennis ball engages more than the muscles of the arm. It requires that we coordinate groups of muscle and stabilize our balance and posture. This takes planning. The primary motor cortex makes your arm move but the planning involves other cortical areas. To make movement purposeful and sophisticated requires other motor regions of the frontal lobes.

The supplementary motor area (SMA) and the premotor area (PMA) are located in the frontal lobe, anterior to the primary motor cortex. Measuring the electrical activity of the SMA and PMA demonstrates that they become active just before the actual execution of a movement. This suggests that the SMA and PMA are important to planning movement. It is believed the PMA may contribute to spatial guidance – directing movement. Amongst other functions, the SMA is thought to coordinate the two sides of the body. How exactly it does so is still being figured out. Structures outside the cerebrum are also involved, particularly where there are complicated coordinated sequences of muscle contraction such as in a game of tennis.

The cerebellum, which sits at the very back and undersurface of the brain, is vital for coordination and procedural (muscle) memory.

A seizure involving the primary motor cortex causes movement of a localized group of muscles. August's seizure manifested as movement, but not movement of a single muscle group or limb – it was frenetic movement involving the whole body. It is not possible to map the SMA and PMA with a small cartoon homunculus. Electrical stimulation of those areas doesn't produce a movement in an isolated body part but rather produces movement of lots of different muscle groups. The importance of the frontal lobes to complex movement implicated them as the cause for August's seizures even without a visible discharge in the EEG.

I discussed the test findings with August. She looked at me hopefully. She seemed happier. Many patients worry that if they come into hospital and we don't see what has been happening to them at home we won't believe them. It seemed that having had two seizures in the unit was improving her mood.

Before I told her that her EEG was normal I told her that I was convinced that her running attacks were epileptic seizures.

"You know one hundred percent?" she asked.

"I am very sure but nothing is one hundred percent."

"The EEG told you it was epilepsy?"

"No, it was the video."

"But my brainwaves told you something?"

"Sadly no. They were unhelpful."

I explained the rationale for the diagnosis to August. She looked alternatively pleased and upset.

"I really wanted the EEG to give you the answer," she said.

"I know, but I really think a diagnosis of epilepsy is inescapable, even without seeing a change in the EEG."

All our talk was about what was wrong and so little about what we could do for her. I sensed that August was desperate for a definitive answer. I think I was responding to that need in her. The diagnosis had wandered from panic attacks to epilepsy and back again many times in the past. She needed to move away from that.

"You believe me?" she said.

"Of course. I don't think anybody ever doubted how bad things are for you, August. It's just that they tried their best and still weren't able to make you better. They didn't know what to do next."

I didn't either. August had tried three epilepsy medications. Any additional drug I might try would have only a minuscule chance of making her seizure-free.

"Am I stuck like this?"

Since her seizures had started, August had not only dropped out of school, she was also unable to work, dependent on her family and largely confined to her home.

"I'll be honest, August, I don't know how to make these seizures go away. But I can see that they must be impossible to live with and I agree with your other doctors that we have to do absolutely everything possible."

That was the start of a long arduous journey for August and me. I began to experiment with new combinations of medication. Each needed to be titrated very slowly upwards in dose. For each drug we tried the process took at least six months.

Occasionally her epilepsy tricked us by appearing to get a little better. That improvement was never sustained. Almost every drug resulted in some negative consequences. Her weight began to fluctuate wildly.

August is statuesque. She stands proud, broad-shouldered, with a striking sort of beauty. Until, that is, she turned up in my clinic having lost stones in weight. It was a shocking consequence of one of her drugs. The tablets had made her seizures a little better but had also caused her to become dangerously underweight. Skin hung from her bones with nothing in between. Her face had become angular and hard.

"I look terrible," she said coldly. I could see she was holding back the tears. "Do I have to keep taking it?"

Of course not. Drugs that eradicate seizures sometimes find new ways to ruin people's lives. I quickly withdrew the drug and replaced it with another. A year later she had regained the weight. Six months later she was overweight. Another drug, another side effect.

"My clothes don't fit. I can't afford to buy new ones. I look terrible."

I feared that August was becoming resigned. I would have hated that for her. She had always been so full of fight. She had been through so much that it was making her tired.

"I'm sick of explaining myself," she once told me.

August's running was uncompromising and dangerous. She ran quickly and unconsciously and with no warning or preamble. She was unaware of cars and curbs and boiling kettles.

Sometimes injuries were the only way she could tell that she had had a seizure. "I stand up and find that I can't walk because my ankle is swollen or my knee is hurting me – and I don't have a clue why." But at least at home she could control her environment. Once August was on a bus when she had an attack. Her only memory is of getting on the bus and taking a seat on the upper deck. Moments later, transported as if by dark magic, she was sitting in a strange living room surrounded by a large multigenerational Asian family.

August had suddenly stood up, run to the door of the bus and jumped off. Whether the bus had been at a stop or had stopped especially for her we could never establish. Once in the street she had run a few yards and then had turned into a garden. From there she had opened a stranger's front door and run into their house. She had awoken sitting on an unfamiliar sofa with a host of bemused faces staring at her.

The family had been kind. They were a lot calmer than I think I would have been had a stranger run through my front door and made themselves welcome. August had left her bag with phone and wallet on the bus so they had helped her to get home. Unbeknownst to August while she was accepting one stranger's help another stranger had taken responsibility for the possessions that she had abandoned on the seat of the bus. That person saw August's abrupt departure and had seen the bag that was left behind. She found August's mother's phone number on her phone and called her. It was frightening for August's mother to receive calls like this. She was relieved to get another call from August shortly after to tell her that she was safe. Later the same day the woman from the bus dropped by August's mother's house to return the lost belongings.

"She hung around for a while after she gave me the bag. Just sort of stood there saying nothing," August's mother told me and laughed, "I think she was waiting for a reward!"

"It was nice of her – I guess she has her own hardships," I said.

"I didn't have anything to give her."

"Of course, I know that, and I'm sure she realized it too."

"People aren't all the worst types," August's mother said.

"No they are not."

With experiences like that August became reluctant to leave her own flat. That made for a dull life – dull, but relatively safe. If only it had stayed that way. After a couple of years August's seizures seemed to get wilier and wilier. At first if she kept all the doors and windows closed her flat would contain her no matter how furiously she ran. She bounced off her own walls at speed. She woke with bruises and scrapes – but at least she had some control. She made sure there were no overly dangerous obstacles and escaped without any broken bones or injuries from which she could not recover.

Things became scary when August learned to unlock doors in her seizures. Abruptly one day she awoke in the street. She knew that her front door had been locked and therefore that she must have unlocked it. I knew it was my job to protect August but I had never seen anybody like her before so I didn't know how to advise her. Her brother began playing with ideas. He wanted to line the walls of her flat with something soft so that she was less likely to hurt herself. That would make her home like a padded cell. It was too difficult and ugly and depressing so we decided against it. Then her brother wondered about a rope that would tether her so that she could not go beyond the confines of her own home. That seemed too dangerous so it was dismissed. In the end an ingenious occupational therapist came up with a solution. She suggested we install a combination lock with a changeable code. Her mother reset it at each visit. August wrote each code down but was careful never to learn it well enough to be able to enter it without thinking. It was an excellent solution and it worked very well. We all breathed a sigh of relief. But epilepsy requires vigilance that no ordinary human being could sustain all of the time. So eventually her seizures caught her with her back turned.

August lived on the third floor of a block of council flats. She had neighbors on either side and also above and below her. Her avoidance of leaving her own flat had started more than a year before. She was tired of waking up alone in a strange place. She was afraid that one day she would run out in front of traffic. A frightening accident seemed to loom over her. She felt calmer at home and only went out if her mother or a friend could take her.

August is fun to be around. She is sociable so it wasn't hard for her to tempt friends to visit her. I can imagine how this looked to the people in the flats around her. She was an apparently able-bodied young woman in her twenties. She didn't work. She stayed at home all day playing music, with a stream of young people coming and going at all hours. Her relationship with some of the neighbors was very tense. There had been a minor altercation with a neighbor who lived above her and who made a lot of noise late at night. August asked him to be quieter but he had not been very accommodating. He didn't know or care to know that August's flat was her whole world and that she was trapped in it.

One day August's mother was visiting when she went outside to put the rubbish out. If her mother was there August was less careful because she knew she had somebody around whom she could absolutely trust. Therefore neither of them worried when August's mother left the front door open for the few moments she would be outside. She could have done this a hundred other hours in the week and it wouldn't have been a problem but at that moment on that day August just happened to have a seizure. Neither August nor her mother can say exactly what happened next. They were both only there for the aftermath. August woke on the pavement outside her own home. A male neighbor was standing right in front of her screaming angrily an inch from her face.

"I felt his spit land on my cheek," August said when she told me about it. "I pushed him to get him away from me."

Piecing things together it seemed that, as August ran, some part of her or her clothing had snagged on the baby buggy the man had been pushing. She had pulled the buggy from his hands and dragged it a few feet until the seizure came to an abrupt end. The neighbor had considered this to be a direct attempt to either take or injure his child. When August woke up she had no idea why he was so angry and, feeling threatened, pushed the man away, which the man later reported to be a physical assault. August had been too upset and frightened to explain herself. The police were called.

She told them she had epilepsy and a few hours later they called me. I called them back. August was in a police cell. She had been arrested for assault and for the attempted abduction of a child. She was not wearing any medic-alert jewelery and they needed corroboration from me. I confirmed that she had epilepsy and that whatever version of events she had told them was likely to be true. I described her seizures and tried to explain how out of her control they were. It was not up to them to decide if August was guilty or not, the policeman told me, but the information I had given was enough to justify that they release August. She would still face charges.

I waited a few hours and then tried to phone August. She didn't answer. Two days later her panicked, upset mother phoned me on her behalf. August was too incensed to talk to anyone. Knowing how proud August was I could imagine this. Her mother begged me to help. I arranged to see both of them at my next clinic.

August walked in the door of the clinic with a scowl on her face. Before I had asked a question she began to talk. She was tired of explaining herself. She did not want to have to ask for forgiveness, yet again, for something that was not her fault. She did not want to have to constantly ask for help.

"I know whatever happened was not your fault," I assured her, "I'm sure the police know that too. But for me to help we need to talk about it."

"It's not the police," her mother told me, "it's the neighbor. The police told him about the epilepsy and he said he doesn't care. He wants her prosecuted."

"They want me out of the flats," August said.

"That's what it is," her mother confirmed.

I promised to support August and I phoned the police again. They said that while they were happy to accept the explanation, the victim had insisted that August be prosecuted so their hands were tied. The police asked me to write a medical report. I dictated an urgent letter and gave it to my secretary to type.

"You sounded really angry," my secretary comforted me when she played it back.

I *was* furious. I did not expect the neighbor to recognize the cause of August's seizure, but having had a full explanation and corroboration of her story it seemed overly cruel to continue insisting that she be punished. August's home was already her prison and now she couldn't bear to look outside in case the neighbor was there. She started to keep her curtains pulled.

"They want her out of the flats," her mother insisted, "and this is their excuse."

The irony was that August also wanted to move to a more suitable home. But she had nowhere else to go. We had been fighting to have her rehoused to a ground-floor flat for years. Both her family and I were terrified that she would eventually come to serious harm falling down the ugly concrete steps in the communal area. Or maybe she would jump from the third-floor balcony.

It took around six months of constant letter-writing before the charges against August were dropped. During that time she rarely went out. She was terrified of bumping into the neighbor again. She began to miss hospital appointments. I started to have telephone consultations with her so that she didn't have to travel if she didn't want to.

At one point I admitted her to hospital for reassessment. I was so worried about her and desperately looking for new solutions. I repeated every test I had done before, hoping beyond hope for some new way forward. August's seizures still happened in clusters. The drugs had barely helped. The new series of tests yielded one piece of rotten fruit.

August's EEG remained normal. Her MRI, however, became abnormal – not because her disease had changed but rather because technology had improved. The scan showed scattered gray dots buried deep in the white matter, somewhere they should not be. These were consistent with a disorder called a neuronal migration disorder.

In the adult brain the cell bodies of neurons make up the cortical layer. Underneath the cortex is the white matter of the brain, which contains the axons that connect one part of the nervous system to the rest. But our brains don't start life that way. They develop almost inside out. The gray matter or neuronal bodies that eventually make the outer layer of the brain start on the inside as neuroblasts. In the first two months after conception, as the fetus is developing, neuroblasts must migrate to their final destination in the outer cortical layer of the brain where they become neurons. Sometimes that migration fails and islands of gray matter are stranded in the wrong place. That can have a variety of effects. In some people it does no harm at all. They live their whole lives not knowing about it. In others it can lead to severe physical and mental disability. Some babies with this problem don't survive. In others it causes epilepsy.

The MRI confirmed that August had epilepsy, if I had ever doubted it. It was bad news. Neuronal migration disorders are genetic conditions. People affected by them can have difficulty sustaining a successful pregnancy and their offspring are at risk of having severe developmental problems. It was another complication to August's life.

It is now more than ten years since I first met August and five years since we found a definitive cause for her running attacks. August cannot have surgery because there are so many stray gray areas in her brain that it would be impossible to say which was responsible for her seizures. She is stuck with medications and they are still proving unhelpful to her. I don't feel that I have helped her at all. Nor have the scans and tests that I subject her to when I am feeling desperate. Despite that failure, August keeps moving forward with her life. Just a different sort of life. One confined within four walls.

"I am starting a cake-making business," she told me recently. "I've sold one already."

She had never been able to go to work so instead she was planning ways that she could work from home. She showed me a picture of a beautiful array of iced wedding and birthday cakes.

"That's amazing," I said. "Have you always wanted to be a baker?"

"No. When I was in school I wanted to be a doctor."

A doctor. I was shocked. I had known August for years and never realized that was a dream of hers. Everything I knew that she did was creative. She draws beautifully. She loves music. She cooks. She makes flowers from sugar paste. She gave me one once and I keep it in a vase on my bookshelf. But – of course – drawing and listening to music and baking are all things that you can do at home. She had adapted her ambitions to fit the life she was given.

"I didn't realize that you were interested in medicine."

"There are a lot of things I wanted to do that I've never done," she told me. "I always wanted to travel but I've never even been on a plane."

Of course she hadn't been out of the country. She could hardly leave her flat. But I'd never thought about it before.

"Where would you go?"

"I want to go everywhere. I want to go to Germany."

"Germany!"

We both started laughing. She wanted to see the whole world – starting with Germany. It seemed so funny in that moment. But where would you start if you'd never been anywhere?

"I've heard it's good there," she said.

"It is. Beautiful countryside and cities. I think you'd love Berlin," I said.

The conversation made me realize I didn't really know August as well as I thought I did. I only knew the version of her that had epilepsy, but there was a girl before who hadn't needed to live in a cage.

"I've enrolled in a cake-making class," she told me. So far she was self-taught. "But my mum has to come with me to all the classes."

She could not risk going alone. Her mother couldn't afford to pay to join the class herself so she sat on the sidelines and watched.

"I think they all thought we were both a bit mad, me with my mum following me everywhere," August told me. "Then I had a seizure and ran out the door. Mum said their mouths were all hanging open. They never saw anything like it. Mum followed me and brought me back. Then she just said to the rest of them, 'That's why I'm here!' That shut them up!"

"Did you explain to the class why you'd run?" I asked August.

"No. Why should I?"

7

RAY

The Brain is deeper than the sea.
　　　　　—Emily Dickinson, "The Brain—is wider
　　　　　　　　　　　than the Sky" (*c.*1862)

Ray and I were sitting side by side in the video telemetry office where the technicians and I watch the videos of people having seizures. Ray had been an inpatient in the monitoring unit for five days. It was his lucky week – he had three seizures in that short time. We were both really pleased. Ray's seizures had never really improved with medication. Some tablets made him a little better but none made any really useful difference. We needed to see his seizures to figure out if there was another way forward for him.

"I'm not sure I want to watch this," Ray said.

"You don't have to. The video is not going anywhere. It'll be here tomorrow. Take a bit longer to think about it."

I video monitor at least six people a week. Yet I can count on one hand those who have asked to see their own video. I've never fully understood this. Most people with seizures only know what happens through the accounts of their families. Those who do watch often seem very embarrassed by what they see. They apologize and explain. A seizure is an involuntary act. It is a result of faulty wiring. It's not the responsibility of the owner. That people remain so awkward in the face of their own seizures suggests that even they cannot always separate the person they are from their disease.

Ray had heard his own seizures described a hundred times. He had requested to see what we had recorded, but he was nervous.

"Are they bad?" he asked me.

"I don't think you'll see anything you're not expecting," I told him.

"I'm worried about the look in my eyes when they happen."

"Honestly, other than what you already know about your seizures there is nothing else to see. I don't think you'll find them too hard to watch."

Ray had had epilepsy for half his life. Two or three times most weeks he lost consciousness for two or three minutes. Less than ten minutes a week in total. It sounds like such a small amount of lost time but it was enough to alter the course of his whole life.

"Go ahead," Ray said and I turned on the screen.

"Tell me how you know that the seizure is starting," I asked as we scrolled through the video looking for the exact moment that this seizure began.

"I've spent years trying to figure out how to explain it. The best I can do is to say it starts with this beautiful feeling. Beautiful but strange. I'm in a lovely, lovely place. Like I'm on a cloud looking down at everybody else."

Ray had thought a lot about it. I thought his description was a clear and generous account of something awful.

"When I have a seizure I usually ask people around me if they're okay," Ray told me. "I think that I feel so well that it makes me worry about other people. I know they can never feel as good as I do."

Ray's seizure was an exchange of ions and neurotransmitters in his brain, a flow of electricity, but he still felt the need to explain his behavior. It seemed to help him to find sense in his unconscious actions.

On the screen we watched the previous day's footage of him sitting in the armchair by his bed. He straightened slightly in his chair, then looked around and searched for something.

"I couldn't remember where the button was," Ray said to me and laughed uncomfortably. We both kept staring at the screen, saw him eventually find the alarm button and press it. He was indicating to the nurse outside that he didn't feel well.

"What do you feel after the aura?" I asked him.

"The lovely feeling is the only bit I really remember."

We looked back at the video. Fifteen seconds passed and nothing further happened. Then the light changed as a nurse opened the door and came into the room. She was responding to Ray's call. She had barely closed the door behind her when Ray stood up like a lightning bolt.

"Fuck off!" he screamed with great vigor.

The nurse was startled and stood still.

"Fuck off!" he shouted again.

Beside me, Ray shifted uncomfortably in his seat. We saw the nurse gather herself.

"She knows it's not my fault," Ray said tentatively.

"Course she does. She sees lots of seizures. She's not fazed," I assured him.

But she looked fazed. As did I, sitting beside him, just watching the replay.

"Remember the word *football,*" the nurse said and began moving towards where Ray stood by his chair.

"I don't remember her saying that," Ray turned to me and then back to the video.

"Are you having a seizure, Ray?" the nurse asked and we both heard him answer clearly.

"I don't know, I might be."

"I don't remember that either," Ray said to me.

Sometimes in a seizure the patient answers as if they are fully aware but the addled brain doesn't retain what has been said. They are on automatic, interacting in a meaningless way. I have had many apparently coherent conversations with people either during or in the immediate aftermath of seizures that the person has not recalled.

"Fuck off, fuck off," Ray shouted again.

Then he gathered a large glob of white spittle in his mouth and spat it at the nurse. She backed away quickly. Ray immediately spat again and the nurse moved further out of his reach.

"I'm not spitting at her," Ray said.

"We know that, don't worry. It's something that happens in seizures."

He continued to shout and spit. I lowered the sound on the video. It was discomfiting. We could still hear the nurse making calming noises in Ray's direction. She did so from the opposite side of the room, keeping her distance. She began asking him to say his name. He didn't answer. She tried asking for his age and his address. He just said, "Are you all right?" as if he thought she was the person who was unwell. "Fuck off, fuck off." The cursing and spitting went on.

As the seizure progressed Ray began to move around the room. We saw him come to a stop standing with his back resting casually against the wall. After a short while the curses wound down and Ray walked back towards his chair. He reached for a plastic cup that sat on his bedside locker and he took a sip of water, then put down the cup again. He turned back towards the door and ambled out of view. The nurse followed him and led him carefully back to his chair and sat him down. Once he was settled she picked up the cup from the table and asked Ray what it was.

"A cup," he said with a puzzled expression. Then she took a pen from her pocket and showed that to him.

"A pen," he said.

"Are you okay, Ray?" she asked.

"Yes," he said, "are you all right?"

"Do you know where you are?"

"Yes."

"Is it over?"

"Yes."

"Do you know that you had a seizure?"

"What? Yes."

Once it was clear that the worst of the seizure had passed the nurse left the room, returning a minute or two later with a colleague. As soon as the nurse reappeared Ray announced with glee, "I think I might have had a seizure."

"I know. I was here with you when it happened," she told him.

"Were you?!" Ray looked surprised.

"Yes, for the whole thing!"

"Did you get it on camera?"

"Yes, we did," she said and he smiled.

I stopped the video and counted how long the whole episode had lasted. Seven seconds from the aura, the lovely feeling, to pressing the alarm button. Twenty more seconds to standing up and for the nurse to run into the room. Forty seconds spitting and shouting. Thirty seconds of looking around and taking a sip of water. Ten more seconds for Ray to leave the room and be led back. Ten more to recover. One minute and fifty-seven seconds from beginning to end.

"It wasn't so bad?" Ray said to me when we had finished watching.

"Not at all bad," I assured him.

*

Ray started having seizures when he was seventeen years old. I met him when he was thirty. He told me how it all started.

"I was studying for A levels when I began to get these odd feelings. They only happened if I worked too hard. I couldn't properly explain what it felt like."

"Did you go to a doctor?"

"No. I told my mum and she said she thought it was epilepsy. Mum likes to think she knows about medical things!" Ray laughed.

In this instance his mother had been quite right. Ray had decided not to listen to her.

"I knew she could be right – but I didn't want to be labeled. I was fine. I was handling it."

Ray's early seizures didn't come with loss of consciousness. They were just a feeling. There was nothing for others to see. They didn't embarrass him or stop him doing things. But then they changed. One day the odd feeling was joined by spitting and shouting and confusion. He ignored it at first.

"I was in denial," he told me. "When it kept happening I told my mother and she took me to the doctor." The doctor referred him to a neurologist and the diagnosis of epilepsy was confirmed. He was started on treatment, but he got no better. The diagnosis was not in doubt. Ray had been referred to my clinic to explore other avenues for treatment. Surgery in particular.

"You've been referred for consideration for surgery. Is that what you want?" I said to Ray the first time we met.

"Not really!" he laughed.

"Okay … but you do want me to look into the possibility for you, at least?"

"I guess there wouldn't be any harm … but I'm not convinced. I want you to look into it even though I don't really want it. Is that okay?"

"Absolutely. We can do the tests and figure out what the odds are that surgery will help. Then you are free to decide to do whatever you prefer."

"My sister is a psychologist, she thinks I should have the surgery. So does my girlfriend."

"It'll be up to you in the end. I certainly agree you should have the tests. Maybe the results of those will make the decision easy for us."

When we first met Ray was having at least one seizure most weeks, and sometimes two or three. There were rare weeks when there were none. Ray is an optimist. Despite their regularity Ray tolerated his seizures well. He once told me that he considers himself to be a very happy person. That is how I see him too. It is not something that many people can say honestly about themselves. Ray would also say that epilepsy has cost him dearly. I also agree with that.

Ray didn't go to university; which he certainly would have done had his seizures not started at such a critical point in his life. Our personalities, our intelligence, our temperaments, our confidence are all at peril if we are affected by a brain disease. The disability of epilepsy is not only seizures. Ray is intelligent and his personality has not been affected, but his memory is very weak.

Memory is very vulnerable to damage from epileptic seizures. They can disrupt the brain in such a way as to stop memories from being laid down or prevent their retrieval. Also the underlying pathology that is causing the seizure may affect memory. Seizures often arise in the temporal lobes. Whatever disease lies there may cause both epilepsy and memory impairment. Finally some epilepsy drugs also affect memory. Almost everybody who has regular seizures complains of memory problems. Ray bears no physical signs of being disabled. The forgetfulness and the volatility of his seizures have challenged him in ways that nobody meeting him casually would guess at.

"It really knocked my confidence," he told me. "I don't go for promotions at work because I worry that I'll be stressed and will start having lots of seizures and then I won't be able to handle it."

Ray once had a seizure at a job interview. He had all but been offered the job when everything shut down and he woke to find the interviewer gone and a glass of water and a box of tissues on the table in front of him.

"It must be really frightening to wake up from a seizure when you're with strangers," I said.

"Actually I prefer having them when I'm on my own or with somebody I'll never meet again. What I hate is when somebody I know sees them. I think I'm lucky to live in London. The streets are full of weirdos behaving strangely! I'm with my people!"

Ray was often with strangers when he had a seizure. Never having been allowed to drive meant that many of his seizures happened on public transport.

"I'm sitting on the Tube and there's one set of people in front of me, then suddenly there's a whole different set of people there," he said.

Because he's a young man, well built and outwardly in good health, I always worry that people will think Ray is just angry and dangerous when he begins to spit and shout. I am particularly scared for him if he has a seizure somewhere like a confined train carriage. I imagine that if he curses at the wrong person that he'll wake to find himself in a fight that he doesn't know he started.

"That's what my sister says too. But it doesn't really happen."

"What's it like to wake up knowing that people saw what they saw?"

"I have this slow dawning realization that something has happened. When I do I just avoid all eye contact. I choose not to engage. I don't explain unless I feel I have to."

"Why would you have to?"

"If somebody gets a bit antsy."

"Does that happen a lot?" I asked him.

"Not really. One time I came round and a man kept saying, what do you want? It felt a bit like he might kick off. I think I'd been asking him if he was all right and he thought he had to answer. I just walked away. That's how it is. I know whoever sees me having a seizure will be gone in five minutes and I'll never see them again, so who cares? One time a woman followed me. She said she knew it was a seizure and that she wanted to make sure I was okay."

I have tried and failed to imagine what waking from a seizure in a public place is like. Falling asleep on the bus is the closest I've got. It's horrible enough to wake up and wonder how you looked when you were asleep. But sleep is an inert thing. Seizures are active. Through my patients' stories I have heard the worst of what can happen. The frightening ones. The sad ones. The near misses. That's what I'm there for. To listen and help when things go wrong.

"Do people get upset with you a lot?" I once asked. I had just assumed they would.

"No. People are nice," Ray said. "They know there's something wrong with me and they try to help."

My patients have been brought into the kitchens and sitting rooms of strangers. People look after their belongings for them.

They drive them home or put them in taxis. They follow them down the street to make sure they are safe. More people want to be kind and helpful than hurtful. Still, you only need one person lacking understanding to make life difficult. Everybody I know who has regular seizures has met at least one of those.

Once Ray was in a bookshop when he had a seizure. He had been standing looking at a shelf when he felt the warning. There was no time to get to a safe place. He blacked out and awoke standing outside on the pavement. A bookshop worker was holding on to his arm. Ray had been seen walking out of the shop with a book in his hand. The police had been called.

"Oh no, I'm so sorry," I said. I felt so bad for Ray.

"It was fine, the police were really nice!"

"Did they arrest you?"

"They put me in the back of the car. I told them that I have epilepsy. They asked me if I was wearing an alert bracelet. But I never wear one."

"Don't you? Why not?"

"Come on!" Ray laughed.

I knew why. Lots of young people struggle with the recommendation to wear medic-alert jewelery. I tell people with epilepsy that they are just like everybody else (and they are), but then I ask them to wear something that tells people they are different.

"So did the police believe you?"

"They asked me what medication I was taking and when I was able to answer without any problem they were cool. They just let me go."

"I assume you didn't go back into the shop?"

"No. But I rang them the next day. I thought I should apologize. The bookshop owner was really angry when I rang. He didn't believe me. I think he was pissed off that the police just let me go."

I really wanted to tell Ray to stop apologizing. It saddened me that he felt he had to do that. I didn't say so. Maybe it helped him to have that interaction after a seizure. It must be reassuring to have the other person say that it didn't bother them. If so it must have been awful to have the bookshop owner do the opposite.

Much as Ray might insist otherwise I think strangers' reactions matter to him. But the bigger minefield was how to tell new people who came into his life about the diagnosis. If he chose not to tell then he risked being found out in some dramatic way. Or if he did tell he could find himself treated differently. The same was true whether in the formal context of work or the more casual one of friendships and relationships. The former in a sense is easier. It is bound by rules.

"Don't tell people your medical history at the interview," I tell my patients. "Once you've been offered the job then there will be an occupational health assessment and that's when you tell."

Ray got a job in the publishing industry when he was in his early twenties. He has worked in the same job for years. He likes it. It's creative. The people are interesting. His colleagues are pleasant and supportive. It doesn't challenge him at all – which is a good thing and a bad thing. Ray has often said to me that he wants to look for another job. Work colleagues have come and gone, negotiating the career ladder. Ray regards himself to be resolutely on the lower rungs of that ladder.

"I told my boss about the seizures as soon as I got the job. He was fine about it but then the first time he actually saw me have one he literally went white. Months later when we knew each other better he apologized for looking so shocked. He said to me that I'd taught him a lot about epilepsy and that he was grateful." Ray looked really pleased to tell me that. I was pleased too.

Ray negotiated the issue of relationships with the same sort of ease that he did the workplace – albeit with a learning curve. One early girlfriend invited him to Sunday lunch with her parents. Sod's law saw him have a seizure halfway through the main course. During it he spat out his food and told her parents to fuck off. When he woke up at the dinner table surrounded by shocked onlookers he couldn't quite negotiate the aftermath.

"Excuse my language but let's get the fuck out of here," he said to his then girlfriend.

He grabbed his jacket and left, his girlfriend hurrying after him.

Now in his thirties Ray has a long-term girlfriend who does a good job of intervening on his behalf when he has a seizure in pubs or restaurants or concerts. People with epilepsy are very dependent on those around them. Families keep them safe and try to mitigate any limitations the disease places on their lives. Ray's girlfriend encourages him to do more than he is always willing to do. Once he refused to go to the wedding ceremony of a friend of hers, despite her and the wedding party wanting him to be there.

"Can you imagine me shouting halfway through the vows? That would look great on the wedding video. Do you take this man? ... FUCK OFF! And a large glob of spit on the back of the father of the groom's head! Don't get me wrong, our friends would have found it hilarious," Ray explained, "but that's the point isn't it? I just didn't want the whole wedding to end up being about me. If I had a seizure everybody would have been talking about me and looking at me."

I suppose it was because every single important part of Ray's life had been dictated by his seizures that he came to see me to talk about surgery.

I reviewed Ray's tests. His MRI brain scan was obstinately normal. Neither his seizures nor his poor memory nor his loss of confidence were reflected on the scan. I admitted him to hospital for further tests. Perhaps his EEG or the psychometric assessment of his cognitive functions would give us more insight into his brain. A while after Ray and I were sitting side by side looking at his seizures. There were two seizures we watched each in turn. They were identical.

"Has it been useful?" Ray asked once we had finished watching.

"Yes, yes it has," I told him.

*

Cursing is a very heterogeneous thing to try to study scientifically. It can be spontaneous and emotion-laden. It can be casual, used for punctuation. If I asked a person in a scanner to curse, it would hardly involve the same brain regions as cursing at a car that cuts us off in traffic.

It is easier to measure the brain's response to hearing a curse. Broca's area of the frontal lobe is involved in language expression

– finding words for us to say and keeping language fluent. There is a separate area for language comprehension – that is called Wernicke's area in the dominant (usually left) temporal lobe. But hearing somebody curse doesn't just activate Wernicke's area, it also results in brain activation in the limbic system and in a structure called the insula. Folded deep inside the brain the insula has many connections with the frontal lobe and the temporal lobe. The insula registers disgust. It reacts to behavior perceived to be outside of acceptable social norms. That might be cursing but it could also be bad grammar. Wernicke's area interprets the words contained in a curse, while the limbic system and insula seem to determine the moral outrage and emotional response to it. The brain processes swear words differently to other words.

Consideration of other medical conditions that are associated with spontaneous swearing (such as Tourette's syndrome) has implicated both the amygdalae and basal ganglia in the production of swear words – the amygdalae because of their role in emotional control and aggression, the basal ganglia (a group of neurons buried deep in the cerebrum) because of their role in impulse control. The basal ganglia are abnormal in a number of conditions including Parkinson's disease. Parkinson's disease does not only cause physical disability manifesting as slowing of movements and tremor, it also has neuropsychiatric features, amongst them excessive impulsivity and the tendency to swear.

Studies of seizure semiology have a lot to say about both swearing and spitting. Both are fairly common features of epileptic seizure. They are believed to be examples of automatism – an automatic release phenomenon that occurs because brain inhibition has been lost. If swearing is automatic disin-hibited speech, that suggests that the language-dominant hemi-sphere is relatively spared in the seizure. Which suggests that the seizure is arising in the non-dominant hemisphere, which for most of us is the right. Spitting is harder to understand but still a recognized seizure symptom. Studies of semiology associate it with right temporal lobe seizures. Why is spitting more likely to arise in seizures of the right temporal lobe than the left? I have no idea. The brain has a mind of its own.

And Ray did something else in his seizure that would seem incidental to the uninitiated. Something ordinary, but not a coincidence because it is yet another well-recognized feature of epileptic seizures. He took a sip of water. There are many such casual behaviors that are useful lateralizing and localizing features in seizures. Nose wiping, coughing, fidgeting, chewing, blinking. Each of these, like taking a drink, are signposts on the seizure semiology map of the brain. In the case of drinking, once again, the right temporal lobe is implicated.

The signs all strongly pointed towards the right temporal lobe but since Ray's MRI scan was normal, many more signs were needed for us to feel sure. But in fact when telling his story Ray had already given another clue. It started "beautifully." Ecstasy is how Ray once described the feeling.

Dostoevsky had epilepsy and was said to experience ecstatic auras. They are rumored to have been the inspiration for his character Prince Myshkin in *The Idiot*. Ecstatic auras have led to epilepsy being linked with religious and mystical experiences.

Neurologists have a long-standing fondness for making retro-spective diagnoses of neurological conditions in historical

Ray

figures. At the age of thirteen Joan of Arc began experiencing hallucinations several times per day. There is a disorder called Geschwind syndrome that associates a hyperreligious personality with temporal lobe seizures. Since both religiosity and hallucinations are potential symptoms of seizures, recent literature has speculated that Joan of Arc's story could all be accounted for by epilepsy. Saint Francis of Assisi's ecstatic vision of a man with wings has been explained in the same way.

But Ray's beautiful strange aura did not have a spiritual or philosophical meaning to me. It was purely anatomical. There is no absolute consensus as to which part of the cortex, when electrically stimulated, will produce such a feeling of elation. However, certain brain areas have been more frequently implicated than others, in particular the medial temporal lobe and insula.

"The right temporal lobe," I had thought when I looked at Ray's video.

I had no MRI corroboration for my suspicion, but I had consistent clinical signs all in agreement with one another. I hoped the EEG would support my theory further.

I minimized the video window and looked at the brainwaves. The first seizure appeared as a wonderful telltale sawtooth pattern ticked along in A2, T4, F8, the triangle sitting over the part of the right temporal lobe that best represents the limbic structures.

There was a second seizure to look at. Oh no. And a third ... what I saw made me utter expletives of my own. These latter two seizures had their discharge exactly where I did not want them to be: A1, T3, F7. The spitting and cursing looked the same each time, but the brainwave abnormalities had the first seizure starting in the right temporal region and the next two on the left.

When I qualified in medicine, somebody like Ray, with no known cause for his epilepsy and a normal CAT scan, would never have been investigated for potential surgery. I age myself when I admit that, back then, EEG was recorded on a long scroll of paper with no possibility of manipulating the data after the fact. A digital recording can be reconfigured endlessly, allowing you to view the data in new arrangements and with new settings. MRI scans have many advantages over CAT scans: they don't involve dangerous radiation so they can safely be repeated several times if needed; the brain can be viewed in many different ways on several occasions. And yet, even with all these technological advances it remains the case that not everybody gets an answer and not all answers are right. Scans and EEGs provide nothing more than guidance. They are all shadow puppets.

There are further uncertainties with the process of tracking seizures that I have been shirking. Both the clinical signposts and the EEG findings are entirely unreliable. There are clinically silent parts of the brain. If seizure starts in one of those it might not produce any overt clinical sign until it arrives at a more clinically noisy part of the brain. Thus what is actually observed in the seizure is more of a representation of where the discharge has spread than where it actually started. Similarly the EEG discharge might have its onset in a hidden part of the brain and what I see in my recording is only its first appearance at the surface. No test alone is trustworthy. To have any faith in them they must all agree. Even then – when all the signs align

– they are wrong about a third of the time. A person like Maya who has a visible lesion on a scan and all tests are concordant with it has only a seventy percent chance of becoming seizure-free through surgery. This suggests that we are misled by the tests at least thirty percent of the time.

When we had finished looking at his video I explained the conflicting results to Ray. While all his seizures looked identical, his EEG had one arising in the right temporal lobe and two others on the left. The appearance of the seizures suggested that the problem was on the right. The MRI didn't break the stalemate.

"So I can't have surgery?" Ray said.

I was not sure that Ray was truly disappointed. I thought maybe he was relieved to have the tests decide, rather than him having to do so.

"It's not a no," I told him.

Because the brain's workings don't show on the surface, doctors are largely restricted to probing the few parameters that can be measured. Many tests do not see the brain as a solid mass as an MRI does, but instead they take an indirect view.

The brain receives fifteen percent of the blood pumped by the heart. A complex arrangement of arteries deliver glucose, oxygen and nutrients to the brain. The brain needs three milliliters of oxygen for every one hundred grams of tissue per minute. The brain is an avid utilizer of glucose. Through looking at the brain's blood supply and glucose use we can gain further information about how healthy it is. An inactive area of brain uses less oxygen and glucose. An inactive area of brain receives less of the blood supply. Scanning for what is missing is like viewing the brain in the negative spaces.

"If you still want to go ahead I have to arrange more tests," I told Ray. "I have an idea where in your brain the problem is but I need a lot more proof."

Would you have a piece of your brain removed after such a vague statement? Ray agreed to keep going with the tests. They were for our information only, we agreed. It was not a commitment to surgery.

I arranged a PET scan – positron emission tomography. In it Ray had a radioactively labeled glucose compound injected into a vein. This traveled through his body to any area that used glucose. The scanner detects the radiation given off by the compound, and creates a color map that distinguishes areas of high glucose use from those of low use. Any area of the brain that shows as a relatively dark patch is thought to be unhealthy tissue. In Ray a dark patch sat on his right temporal lobe.

Next I sent Ray to have a SPECT scan, which stands for single-photon emission computerized tomography. SPECT uses a radioactively labeled tracer to show where blood flows. The SPECT was done when Ray was well, and another was timed to be done when Ray was having a seizure. That is just as difficult as it sounds. It involves a nurse staying never more than a foot from the patient's bedside, plus a great deal of luck. She is there to inject the tracer as soon as the aura begins. During a seizure blood rushes to the electrically active part of the brain. SPECT would therefore follow the blood as it rushed to the site of the seizure onset in Ray. Ray's SPECT implicated an area in the right temporal lobe.

Short of opening his skull, Ray's brain had been looked at every way we knew how: structurally, electrically, glucose utilization, blood flow. I discussed the details of Ray's tests at the hospital's multidisciplinary meeting and got the views of the full team. Then I met with Ray to let him know what had been said. Ray had come with Rona, his girlfriend.

"So ... looking at the video of your seizure it looks very much to me as if the problem is arising in the right temporal lobe. The other doctors agreed with that view," I said.

"Okay ..." Ray was hesitant.

"And the PET and SPECT scans have supported that hypothesis. They both showed abnormalities in the right temporal lobe."

"Good."

"Your memory tests show that your visual memory is very weak. So that part of your right temporal lobe isn't doing too much. You knew that of course."

"Yes, my memory has always been terrible."

"Your verbal memory is strong. That's good too – it also pushes us toward implicating the right temporal lobe."

"Okay ..." Ray said, "I'm waiting for the 'but'!"

"However – the EEG has shown an abnormality in the right temporal lobe and in the left temporal lobe in different seizures. And the MRI scan is totally normal."

"Which means a no or a maybe?" Rona asked.

"It means that we have a very strong suspicion that the seizures are from the right temporal lobe but cannot be one hundred percent sure. Also the temporal lobe is a big place. The surgeon won't remove the whole temporal lobe, only part of it. When the scan is normal it is hard to say which part that should be."

"So a dead end?" Ray asked.

I hesitated.

"No ... not a dead end ... but even more needs to be done to hone this down. To take out a bit of brain that looks healthy on scan is a big decision. You wouldn't want that and the surgeon wouldn't want to do it. We need more evidence to say it's the right thing. To get that evidence we would need you to have an intracranial EEG. So far we have narrowed down the target area for surgery. But we need both to confirm that those assumptions are correct and narrow the target to a smaller section of temporal lobe. That is done by opening part of the skull and placing a limited number of sterile electrodes directly on the surface of the right temporal lobe. Then we wait for yet another seizure. It's like having video telemetry again but this time we record directly from the brain with no interference from muscles and the skull. The surgeon can put electrodes on bits of the brain that are hidden from the view of scalp electrodes."

"An operation before the operation?"

"Exactly, yes."

"What are the risks of putting electrodes on my brain?"

"We are placing a foreign object on the surface of the brain, so that risks infection. There is also the risk of having a stroke. If either of these occur that would obviously be very serious. Of course the surgeons who do this operation do it very regularly. They are very experienced so risks of serious complications are low. But still there are no guarantees."

"But if it's successful it will lead to surgery?" Rona asked.

"Only if it tells us what we need to know. If it shows an electrical discharge that is very discrete and in the right temporal lobe then, assuming you want it, we can go to surgery. If it doesn't show a clear focus in the right temporal lobe then it will mean surgery is not currently possible."

If I thought that Ray looked dubious before, that look was now magnified.

"I have an operation to plan for an operation that might never happen?"

"Yes. The multidisciplinary team thought that if you had the intracranial EEG study there would be a seventy percent chance that it would lead to surgery. Then if they did the surgery that would come with somewhere in the region of a forty percent chance that you would become free of seizure. So we are looking at roughly a thirty percent chance that the surgeon will cure your epilepsy. These are all estimates, of course."

"So a thirty percent chance I go through the intracranial test and don't even get to the surgical stage?"

"I'm afraid so."

"And one third chance I'm cured at the end of it all?"

"Yes ..."

"I get a buzz cut to one side of my hair ... and a bit of brain chopped out ... a week in hospital and months off work ... and a two out of three chance that I won't be any better off?"

"That's about the size of it. The haircut is for free ..."

"Ha ha. Mmm. I could also end up worse off ... I could have a stroke that turns me into a vegetable."

"Big risks exist, but they're unlikely. The most likely negative outcome is that it just wouldn't work. But remember why we are offering this in the first place. You're having seizures every week. You haven't come to harm in any of them but if this continues you could come to harm in the future. Also each successive seizure can worsen the memory. Or you could develop new seizures from a different place in the brain. If that happened surgery would no longer be possible. That's why we do this now when you're well. I don't want to wait until you get worse and then it's too late."

Ray looked at Rona. "I don't know how I'm supposed to make this decision," he said. "I feel too well to go through this."

"What do you think Rona?" I asked.

"I think he should have the intracranial thing."

"I know my sister thinks I should have the surgery," Ray said. "But my mother thinks I shouldn't. I feel too well to have it. I've always just got on with it. I don't know that my epilepsy is bad enough for this."

"I think three seizures a week is quite bad ..."

Ray's perception of how things are for him has always been more optimistic than mine. He is a happy person. Perhaps it was wiser not to mess with that.

"What do you think I should do?"

I didn't know. I was wavering.

"I suppose you're not supposed to tell me what to do ..." Ray added when I hesitated.

It was Ray's choice but I had the power to influence him as strongly as I wished.

"I think you should have the intracranial study at least. Then you know what's possible," I finally said.

"Can you put me in touch with somebody who's already had this done? It would help me decide."

Of course I could, but who? I could pick a success or a failure. There is no uniform experience.

*

Gabriel developed epilepsy when he was in his twenties. He went through the same process as Ray when he was in his mid-forties. At the time he was a sales manager for a large firm. I always assumed he must be very good at his job. He couldn't drive due to his epilepsy so he was assigned to locations in central London so that he could travel by public transport to meet clients. This suggested to me that his company valued him. Gabriel was married with three children. Essentially his life was as normal as any person's could be if they had to live with seizures. That was precisely why Gabriel decided to opt for an intracranial EEG study. He worried his epilepsy would get worse and he would lose everything. He saw surgery as an opportunity to prevent that from happening.

He had frontal lobe seizures. He had had his first driving his car. The attacks caused wild flailing movements in all four limbs. It was as if each arm and leg had a life of its own and was vying to escape the body they were attached to. They twisted and grabbed at things outside Gabriel's control. The first attack caused Gabriel to crash his car. He never drove again. They were hyperkinetic seizures typical of the frontal lobe. Gabriel had them every week. They could easily have been very life-limiting but Gabriel simply didn't allow them to be.

Gabriel's MRI scan was normal and his other tests could only hone down the location of the seizures to a large area of the right frontal lobe. The EEG study helped identify a circumscribed area where the seizures were most likely to be arising. To the naked eye that bit of brain looked normal. On the scan it looked normal. Its electrical pattern suggested otherwise. It was removed. Gabriel's seizures got better. To date they have never recurred.

Three months after surgery Gabriel became depressed. He had some psychotic symptoms and became prone to paranoia. His thinking became irrational. He tried going back to work to regain control and normality in his life. His behavior became more and more erratic. Clients remarked to his manager that his behavior was bizarre. His wife began to complain that he was hard to live with. His moods were so unpredictable that his own children didn't want to be alone with him. One by one the stitches of Gabriel's life were unpicked. He was forced to take more time off work. Confinement at home made him more and more depressed. His volatility alienated his family. A year after surgery Gabriel's wife asked him to move out. For the children's sake. Shortly before, he had been made redundant.

The surgeon's part of the job was a major success. The seizures had stopped. But some aspect of that was proving too much for Gabriel's brain. Depression is a distinct risk after brain surgery. Seeing the psychiatrists before surgery helps prepare for this eventuality but won't prevent it. Gabriel found himself hospitalized with post-surgical psychiatric complications. He spent three months on a psychiatry ward. He recovered, but not enough to get his life back. He is divorced. He lives alone. He sees his children sporadically. He is unemployed. Surgery is not only about surviving the operation. Being seizure-free has not yet benefited Gabriel. I can only hope that there is still time for him to reap the benefits. Had Gabriel not had surgery his life would not have suffered an overnight change and he could have gone on happily as he was. Or maybe he wouldn't. Maybe he would have come to serious, even fatal harm during a seizure. Maybe surgery saved his life. Maybe his memory would have declined so much that he would have lost his job anyway. There is no way of knowing if any of these difficult decisions were right or wrong.

Susan also had frontal lobe seizures with a normal scan. She went through the full range of tests to investigate if surgery would cure her, and was told that she had a twenty to thirty percent chance that surgery would make her substantially better. It is a testament to how desperate she was that she agreed to go ahead.

Susan's seizures caused her to jump on the spot. The movements looked very much like an addled, drug-fueled dance. A rhythmless jig. Afterwards she lay on the ground and performed pelvic thrusting movements. Each attack lasted less than a minute. More than once she was nearly arrested. Usually she was accused of being intoxicated. I gave her a letter that she carried everywhere to explain her medical condition. Eventually, after she had woken surrounded by confused observers enough times, she stopped going out without a member of her family.

Susan was in her thirties when she had surgery. She regarded it as her last chance of a normal existence. That brave decision proved to be the right one. Six months after surgery she sent me a picture of herself abseiling down a rock face. I was furious.

Had she asked me if she could go abseiling so soon after surgery I would most certainly have said no.

"There was lots of safety equipment. They knew about my epilepsy and they said it was okay," she told me, laughing when I let her know that she had terrified me with her new daredevil life. She's never looked back. She's seizure-free now and endlessly grateful to the surgeon who changed her life, hopefully forever.

*

Who would I put Ray in touch with?

"I think the pre-surgical counselor is the best person to talk to about surgery," I told Ray. "I can introduce you to other patients but your experience could be very different to theirs. It could be misleading."

"Okay."

"Okay – you'd like to talk to the counselor first?"

"Okay, I'd like to talk to the counselor and go ahead with the intracranial study."

Rona nodded her head and put a hand on Ray's arm.

"I'll write to the surgeon and let her know."

Three months later Ray emailed me. *I don't want the surgery. My life is good and I can live with the seizures.* I found myself unexpectedly relieved. I wasn't sure what he should do but I found myself joining him in having more trust in the status quo. *Feel free to change your mind at any time,* I replied. *Every year there are new advances so it may well be that we can do all of this much more precisely and much more safely in five years.*

Technology has been advantageous to the practice of clinical neurology, but it is still very far away from fulfilling all the promises it has made. In the field of neurosurgery, however, technology has been much more transformative. And the advances keep coming. There is reason to be hopeful for Ray. Very soon there may be minimally invasive procedures to remove bits of brain with much less risk. In a few years Ray may be able to have laser surgery, negating the need for a major open operation. Or a stimulating electrode may be placed in his temporal lobe to quiet the electrical activity of his seizures.

But we are not there yet.

8

LENNY

The Brain is just the weight of God.

> —Emily Dickinson, "The Brain—is wider
> than the Sky" (c.1862)

Lenny's journey to the diagnosis of epilepsy was not especially unusual. In his early teens he developed episodes a little like panic attacks. Every now and then, out of the blue, he became anxious. Sometimes it evolved until he felt he couldn't catch his breath. It came with a feeling of doom. He didn't lose consciousness or collapse.

Lenny discussed the problem with his doctor, who diagnosed an anxiety disorder. Every teenager has something to be anxious about and Lenny was no different. School was a struggle for him at times. He was a quiet and thoughtful person who didn't always feel comfortable in a boisterous environment. His symptoms were attributed to that. He began to see the school psychologist. This helped him a great deal. It didn't get rid of all the symptoms but made it easier for him to live with them.

Then when Lenny was sixteen he came home one evening and appeared to be behaving very strangely. He was laughing inappropriately. Not everything he said made sense. He had torn his trousers but couldn't remember how it had happened. His mother was convinced he was drunk or had taken drugs. Lenny absolutely denied it. He told his mother he had been skateboarding in the park. He didn't remember falling but maybe he had.

"Where's your skateboard?" his mother asked him.

Lenny didn't have it with him. He must have left it in the park.

"Who were you with?" she asked.

Lenny told her he had been on his own. She drove him to the park to see if his skateboard was there. It was. When Lenny's father arrived home they discussed whether or not they should take him to the doctor. His mother wondered if he had fallen and hit his head, but other than the torn clothing there was no sign of injury. In the end they didn't do anything. After an hour or so Lenny seemed better. Overnight his mother checked on him in his sleep and he was fine so the mystery was left unsolved.

The diagnosis of epilepsy depended on something else happening to draw attention to the real problem. Only when the electrical discharge spread to the whole brain and caused a generalized tonic clonic seizure did someone realize that the panic attacks had never actually been panic attacks. They had each been the result of small seizures confined to the limbic system. The limbic system, essential to the control of and expression of emotion, can generate palpitations and sweating and all the physical features of panic.

Lenny's focal seizures, masquerading as anxiety, had slowly evolved over time. After a year or so he started to notice that something new followed the initial anxious feeling. It could last for several minutes.

"I call it my 'maybe I will, or maybe I won't' feeling," Lenny said. "It's like I can see all the possibilities and I just can't choose. Every decision seems equally right or equally wrong. It's a horrible feeling. I hate it."

When Lenny first started getting this sense of insurmountable indecision he didn't tell anybody. He had come to simply accept that he was an anxious person.

He was at school when the first generalized seizure hit him. He had the usual sense of uncertainty and worry in the moments leading up to it. He didn't tell anybody about this because he was so used to it. This time, however, it didn't stop. He began to feel very nauseous. The room turned black and he collapsed. Classmates said that his body stiffened and he slid from his chair under his desk. The teacher ran to help and found him completely unresponsive. His body was stiff and his breathing was very loud, almost like a snore. His lips had turned blue. The teacher thought he was dying and attempted to give him mouth-to-mouth resuscitation. Before the teacher had properly started, Lenny took one loud, long snorting intake of breath and began to recover. Within a couple of minutes he was sitting up asking people what had happened.

Lenny went to the casualty department and shortly afterwards saw a neurologist. Once he had given his account of the odd warning that he had in advance of the attack that was enough. His description fit beautifully with a seizure aura originating in the temporal lobe and he was given a diagnosis of epilepsy. An EEG showed spike discharges in his right temporal lobe, confirming the diagnosis. Lenny was offered epilepsy medication.

"I refused the tablets the first time they suggested them," Lenny told me years later. "I'd only collapsed once and I felt fine. I was used to the panic attacks. I said I didn't need the tablets. I think I was in denial. Taking the tablets meant admitting I had epilepsy and I didn't want to do that."

Eventually Lenny's seizures gave him no choice. When he had collapsed three more times he relented and took the medication.

"I was glad in the end. The panic attacks got much better. I still had them but not so much."

The anxious spells went from one a week to one every two months. About fifty percent of them led to a convulsion, and over three years, Lenny collapsed eight times. Two additional epilepsy drugs were trialed. When they didn't work a decision was made to investigate Lenny for the possibility of surgery. Since seizures in the temporal lobe often produce anxious feelings, his symptoms, combined with the EEG abnormality, gave a working theory that he had right temporal lobe epilepsy. His MRI brain scan was normal, but he was an otherwise healthy young man, so in many ways he was a potentially good surgical candidate.

I brought Lenny into hospital to video his attacks. It took three admissions for us to witness a collapse. During the first stay he had had an anxious warning feeling, but nothing more. The EEG didn't change. Very subtle small seizures are not always detectable on scalp EEG. We needed a bigger attack. The seizures were often weeks apart so were bound to be hard to capture. On the third admission Lenny stayed in hospital for two weeks and his epilepsy drug doses were reduced. We hoped this would sway the odds in his favor.

We were rewarded when on day five Lenny finally collapsed. It happened in the evening and the nurses reported it to me the next day. I looked at the video.

Lenny was sitting back on the bed playing with his mobile phone. He abruptly stopped what he was doing and put a hand to his stomach. He looked uncomfortable. He nodded his head and then looked around. He lifted one hand and waved at the camera to let the nurse watching him know that he wasn't feeling well. He realized a moment later that he was supposed to press the alarm button and did so. He kept rubbing his stomach with his right hand and nodding. A nurse came into the room.

"Are you having a seizure?" she called out as she came through the door.

"Yeah, a small one I think. I just feel a bit odd. That's what it's like."

The nurse began to ask him to identify objects in the room and to give his name and address. At first he answered easily, but then halfway through saying his address he exclaimed, "Wo … aah," and seemed to take a big breath. His eyes rolled upwards and flickered slightly. His mouth hung open limply. He lost consciousness and fell back against his pillows. He jerked once or twice.

"That's not a generalized seizure," I said to the technician watching the screen over my shoulder.

"Nope, it's weird," she said.

We were looking for a generalized tonic clonic seizure – prominent stiffness and rhythmic jerking. Neither of these signs were very convincing in Lenny. His body was not stiff; it was limp. Nor were there any more than one or two convulsive movements. Mostly he lay still, with long gaps between gasping breaths.

I minimized the video window to look at the EEG tracing which was hidden behind it. Usually in a seizure we see a rhythmic electrical discharge. Lenny's EEG showed no excess of electrical activity. The brainwaves were almost flat. There was not an excess of electrical activity but rather a paucity of it.

"Where's the ECG gone?" I wondered aloud. "Is it something technical?" I asked the technician.

Perhaps the headbox lead had broken. Or had become unplugged.

A heart tracing (ECG – electrocardiogram) runs continuously alongside the brainwave recording. Lenny's line, that should have shown his heartbeat, had flattened. His brainwaves were disappearing too.

I scrolled back a few pages to the point where Lenny first reported feeling unwell. The EEG and ECG looked quite normal at that point. I pressed play and let the EEG play across the screen. As Lenny waved at the camera, both were still normal. Just a few seconds before he lost awareness, as he rubbed his stomach, the brainwaves changed. The rhythmic pattern of a temporal lobe seizure appeared reassuringly in the right temporal region. Diagnosis confirmed. Seconds later there was something new. The heart tracing began to look different. First it changed shape, then it slowed. Lenny's heart rate before the seizure was eighty-five beats per minute. Shortly after the seizure discharge appeared, his heart rate dropped dramatically to forty beats per minute. And then, as he sunk into the pillows and the nurse tended to him, it stopped. Completely. He flatlined.

As usual I was watching an event that had happened the day before. On another screen behind me, the real-time Lenny was flicking through the TV channels. I knew that in the video he would wake up and yet, waiting for his heart to beat again, I could feel myself holding my breath in fear. For twenty-five seconds there was no heartbeat. Lenny was as good as in cardiac arrest. The nurse caring for him didn't know. She had no reason to. She was primed to respond to epileptic seizures, not heart problems. If this had been a cardiac unit she would be pounding his chest by now. Instead she rolled Lenny onto his side to protect his breathing and waited for him to wake up spontaneously. He did.

When he first reported feeling anxious and then lost consciousness, that was a result of the electrical discharge in his brain. It was when this discharge moved to the control center for the autonomic nervous system that things changed. The heart slowed, then stopped. Blood pressure fell and his brain was deprived of oxygen. That caused Lenny's brainwaves to flatten and potentially to disappear. The strange evolution of his seizure was due to the fact that his brain was lacking oxygen. Our brains need only be deprived of oxygen for three minutes for permanent brain damage to occur. Thankfully for Lenny his heart kicked in and his blood flow set off again after a frightening thirty seconds. This was a cardiac-induced faint. Lenny did not have a problem with his heart per se. He had a brain disease, but the symptoms presented in his heart.

The autonomic nervous system is a confluence of peripheral nerves that serve the internal organs, blood vessels, skin, tear ducts and pupils. The nerves that go to the heart, lungs and blood vessels help regulate the level of oxygen and carbon dioxide in our blood, and thus the amount of oxygen reaching our brain. It speeds and slows the heart in response to physical and emotional triggers. The autonomic nervous system also makes tears come from our eyes and sweat from our glands. It maintains control over the movement of the bowel, the emptying of the bladder, respiratory rate and sexual arousal. It influences heart rate, breathing rate and blood pressure. The hypothalamus is the brain's control center for the autonomic nervous system. The hypothalamus has close connections to the amygdala, the hippocampus and the olfactory cortex. Each of these, when damaged, is a frequent source of seizures.

The involvement of autonomic control in a seizure has two potentially dangerous consequences. The first is that the heart stops. The second is that breathing control is severely affected. In response to pain or lack of oxygen, the autonomic centers can tell your breathing to speed up. It can also tell it to slow down. You don't think about your breathing because your autonomic nervous system is thinking about it for you. If you try to hold your breath it will at some point override your conscious decision and force you to breathe again. In a seizure the system may fail and cause either hyperventilation (breathing too rapidly) or hypoventilation (breathing too slowly). Both hypoventilation and apnea – a complete cessation of breathing – can cause a dangerous fall in blood oxygenation, making them life-threatening features of epilepsy.

Any sort of seizure can ultimately involve the autonomic center, but it is particularly common with seizures arising in the limbic system and the insula (that folded piece of brain that has many circuits connecting it to the temporal and frontal lobes). Neurostimulation of the insula has been shown to alter heart rate. Seizures in both the temporal lobe and frontal lobe can involve the insula and then spread to the limbic structures, the amygdala and hippocampus. The limbic system communicates directly with the hypothalamus which facilitates the subsequent autonomic response.

Some degree of heart-rate change – whether speeding up or slowing down – is very common in seizures. Thankfully it is very rarely fatal. Seizures are mostly finite and brief, and the heart recovers when the brain does.

Lenny's seizure sung of the temporal lobe. He had anxiety attacks of the limbic system, then a feeling of doom and indecision, followed by cardiac involvement. The problem was that I had no way of knowing if Lenny's heart stopped with every seizure. Perhaps this seizure was worse than usual because we had reduced the strength of his drugs to improve our chances of seeing one in the video telemetry unit. After all Lenny had had epilepsy for many years and had not yet come to harm. He always woke up. If his heart had ever stopped before it hadn't lasted long enough to harm him. Nevertheless I referred Lenny to a cardiologist. They put in a pacemaker. It wouldn't stop Lenny's seizures or make any difference to his anxious auras, but it would mean that in the next seizure he would not have a cardiac arrest.

Lenny didn't have surgery in the end. After the pacemaker was put in his collapses dwindled. I am tempted to say that all of the previous blackouts were due to brain-disease-induced cardiac arrhythmia, but I can't be sure. The anxiety attacks have continued but Lenny prefers to live with those than go ahead with surgery. For now at least.

Awake or asleep our brains are never fully resting. They are always engaged in maintaining the normal rhythms of life. It is very easy to forget the importance of the brain's role in the unconscious work of all our internal organs. The bowel and bladder, the heart and lungs, endocrine glands, sexual organs, sweat glands, skin and pupils are all under the jurisdiction of the autonomic nervous system. Any one of these can be involved in a seizure. Facial flushing, goosebumps, vomiting, stomach gurgling, belching, sweating, pupillary dilatation, palpitations, incontinence and sexual arousal are all symptoms of brain disease.

*

Tim's mother, Maureen, was sitting at the kitchen table when the doorbell rang. She was not expecting anybody. She assumed it was charity workers collecting old clothes, or somebody looking for the family next door. There was constant confusion between the addresses, one being 39 and the other 39A.

Maureen had been cleaning the house. She hadn't even brushed her hair that morning. She didn't want to answer. Instead she went into the living room and peeped out from behind the curtain. When she saw two policemen standing there she got a fright. Then she saw a shaggy-haired youth standing between them and felt better. He was a friend of the neighbor, she assumed. They had a couple of sons his age still at home. Always stumbling up the path at all hours. Although they weren't usually ones to get into serious trouble, as far as she knew. The boy stood between the police officers with his chin on his chest. Ashamed of something, Maureen told herself.

Maureen let the curtain fall back into place, but didn't go back to the kitchen straightaway. She waited to see if the callers would realize their mistake. The doorbell rang a second time. With it came a short row of knocks. Her husband was at work. With both children at college she was alone. She hated answering the door when she wasn't presentable. They gave her no choice when they knocked again. She opened the door halfway and looked out from behind it. The young man lifted his head. Seeing his face, Maureen thought she recognized him but out of context she couldn't place him. She didn't immediately say hello.

"Mrs. Dolan?" one of the policemen said.

Maureen glanced back at the boy. He was upset about something, she could tell. Something clicked in Maureen's brain. She knew who this boy was. She would have given anything to close the door then and there.

Maureen and Jack were from Ireland. They had moved to England for work and settled there. They were both in their late twenties when their first son, Sean, was born. Tim came along three years later. The two boys were close when they were very young, but had grown apart as they got older. Sean was a bookish academic boy, while Tim was more interested in sport and socializing. Both boys were clever but Sean could apply himself and Tim could not. Maureen maintained that Tim was the brighter of the two but lacked the discipline to get the exam marks that her older son was able to achieve.

Maureen and Sean were particularly close, as were Tim and his father. Tim's gregarious nature and love of rugby appealed to Jack. He enjoyed going to his son's rugby matches. Tim was not an exceptional player but he was good enough to play for the local club's first team. They were not a great team, Maureen told me, but he was very proud to play for them. "An enthusiastic amateur" his father called him, after he had finally accepted that his son would never have a career as a professional player.

Maureen suspected that Tim's love of the rugby club wasn't just about the game. He talked about the parties almost as much as he did the sport. She could tell that she would need to keep a much closer eye on her younger son than she ever had on Sean.

The first seizure Tim ever had happened the morning after he had played one of the biggest matches of the year. Tim had just turned sixteen and Sean had already moved out to go to university. It had been a home match played at the ground a few miles from where the family lived. After the game the team had all gone back to one of the players' houses. It was more usual for them to stay on at the rugby club, which meant that Maureen or Jack would collect Tim and bring him home at around midnight. But since he was at a house party a fifteen-minute walk from their house he was trusted to get home alone. As a consequence, Maureen lay in bed all night worrying until she heard her son stumbling through the front door at five in the morning. Jack had to stop her from getting straight out of bed to tell him off, reminding her that in the small town in

Ireland where they had grown up they had each started drinking and staying out long before they were sixteen. If anything, Jack said, their sons' city upbringing had slowed everything down for them. Tim was almost always required to come home early on public transport or with the taxi of Mum and Dad. Maureen reluctantly agreed to let Tim get away with it this time.

"He'll have left home in two years and then you'll be stuck with just me," her husband reminded her. "You don't want to drive him away early."

At midday the next day there was no sign of Tim. Maureen peeped into his room at one stage and heard her son grunt in his sleep.

"The smell in that room. He drank a lot last night, I'd say," she told her husband.

At one in the afternoon when Sunday lunch was ready Maureen decided to get Tim out of bed. She opened his door and called his name but he didn't answer. She went in and threw open the curtains and opened a window to let the fresh air in. When she turned to look at Tim he was lying on his side snoring. It was only as she got closer that she saw a trail of blood on his pillow and more blood dried around his mouth. She shook him but he didn't respond. His eyes were half open and the snoring continued. Maureen shook her son harder. She began to panic. She called Jack and when neither could wake him they called an ambulance.

Tim was taken to the nearest hospital. His parents followed behind in their car. They ran in the door of the casualty department about ten minutes after their son had been admitted. A nurse told them Tim was in the resuscitation room.

For five tense minutes they waited in the family area for news before a nurse came and told them that Tim was awake. He had started to come round in the ambulance and while they still didn't know what was wrong with him the immediate danger was past. Maureen and Jack were allowed to see him. He was groggy and disorientated but much better than when they last saw him.

"For the size of him he still looked like a little boy," Maureen later described how it had felt to see her son vulnerable on a hospital trolley.

Tim didn't get a definite diagnosis on that first trip to the casualty department. Doctors insisted on doing a thorough drug and alcohol screen. Although Tim admitted drinking a lot the night before, his alcohol levels were already zero. His drug screen was clear. A brain scan was also normal. Tim knew nothing about what had happened, but neither could he remember several hours from the night before. Despite Tim's objection, Maureen insisted on speaking to some of the other boys at the party. Nobody had noticed anything strange about Tim's behavior. They all denied having taken drugs.

Tim was sent to an epilepsy clinic on the off chance that he could have had a seizure. Even though the tests were normal, the neurologist still considered a seizure to be a strong possibility. But as it was an isolated event he didn't feel that anything further needed to be done.

Three months later Tim had a witnessed seizure. It was a Sunday morning once again. Tim had been out the night before but his mother had collected him from the rugby club at midnight. Maureen and Jack's bedroom shared a wall with Tim's room. Maureen was woken that morning by a loud strangled scream. She ran to her son and found him rigid and shaking in bed. His eyes were open. His face was set in a grimace. Blood trailed down his cheek. His lips were blue and he didn't appear to be breathing. Jack ran into the room and ran straight out again to call an ambulance. By the time the paramedics arrived, which was only a few minutes later, Tim had stopped shaking. He had been unrousable but was beginning to come round.

"He fought the paramedic," his mother recalled. "He kept trying to push everybody away. I don't think he knew what was happening."

Tim was taken to casualty. Within three hours he had fully recovered and was able to go home. He met the neurologist again. He concluded that Tim had epilepsy. The second seizure had proven the cause of the first. Tim needed to be on treatment to prevent a third.

Maureen had been very upset about the diagnosis but Tim was less so. Only when he was advised to cut down drastically on late nights and alcohol did he get at all upset.

The type of epilepsy that teenagers get is often very sensitive to both sleep deprivation and excessive alcohol. It is not uncommon for the first seizure to happen once a teen has started drinking. Certainly both of Tim's attacks fitted with this pattern. His doctor told him he could only drink small amounts of alcohol from then on. Tim made the doctor explain what he considered a small amount of alcohol.

"One drink," the doctor said.

"What's the point of having one drink?" Tim replied, much to the embarrassment of his mother.

"Are you not able to have a good time without drinking?" she said.

Despite his objections Tim complied with the doctor's instructions. The promise of getting a driving license helped. Tim's father told him that if he didn't drink and didn't have any more seizures then he would buy him a secondhand car. Tim had to be seizure-free for a year to apply for a driving license.

Unfortunately Tim's epilepsy didn't comply with his goal to have a car before he went to university. The first epilepsy drug he took seemed to make him worse, not better. He had two further seizures in rapid succession. Like the first two they happened in bed shortly before Tim was due to get up.

"Also he has become very clumsy," Maureen told the doctor. "He drops things and spills things."

Tim had said he felt twitchy all the time. In the morning in particular he spilled his orange juice or knocked things over on the breakfast table.

It was clear that Tim was having generalized convulsions, but not which type – generalized epilepsy from the start or a focal seizure where the discharge spread to become generalized. The distinction matters because the medications are slightly different.

Nobody had seen Tim's seizures from the beginning and that is where all the big clues reside. The new clumsiness that Tim and his family noticed cleared that mystery. Generalized seizures come in different sorts. Convulsions are those most talked about but two other common possibilities are blank spells (absences) and myoclonic jerks. In a myoclonic jerk there is a very brief electrical burst in the brain accompanied by a lightning-quick muscle jerk. Those muscle jerks are so fast that they can be hard to spot but still cause the affected person to drop whatever they are holding. Myoclonic jerks tend to happen first thing in the morning. They are just as Tim and his mother described.

Tim therefore had two sorts of seizures – convulsions and myoclonic jerks. This meant he had a very specific epilepsy syndrome that begins in the teen years called juvenile myoclonic epilepsy. It can be made worse by some medications, as Tim was experiencing.

This discovery was a relief to Tim and his family. Now they knew why he wasn't getting better, something could be done about it. His medication was changed. The jerks stopped. Tim had one more seizure four months later. His drug dose was increased as a result and he seemed to go into remission.

Tim got better. He sat his A levels. They didn't go well. He had been offered a place on a degree course in business and language studies but he didn't get the necessary grades.

"May God forgive me, but I remember I was delighted," his mother told me. "I didn't think he was ready to go. He was much more immature than Sean was at that age and, with the epilepsy, I did worry."

Tim ended up staying at home a year longer than he wanted. He re-sat his A levels and reapplied for university. This time he got in.

"He liked being looked after and college was easy enough, so he enjoyed that year once he got used to the idea of it," Maureen said.

"Except his friends had moved on," his father said.

"Yes," Maureen said quietly, lost in thought, remembering.

In the year that Tim stayed at home waiting to go to university he did not have any seizures. He got his driving license and his father bought the car he had promised. By the time he packed his car and moved out Tim had not had a seizure for over eighteen months.

"When they are growing up you can't wait for them to leave and then when it comes around you're not sure you want it after all," Maureen said.

Dropping Sean off at the university halls of residence four years earlier had been easy. But Tim was still at home then, so the house hadn't seemed so empty. And Sean didn't have epilepsy.

"Young people go a bit mad when they go to university and move out of home," Maureen worried at the time, "there's so much temptation out there."

Tim had had to promise sincerely that he would not drink too much and he would sleep as much as he needed. And take his tablets regularly. And always remember to fill his prescription in plenty of time. And eat properly. And not drink and drive. And not overpack the car with friends.

Whether Tim obeyed all the rules or only some of them is not known. But he did have a seizure in the first term. A friend who did not have a room in the halls of residence was sleeping on Tim's floor after a night out. He was woken by Tim's loud cry and found him convulsing in bed. Although Tim was already awake by the time the paramedics arrived, one of his flatmates rang his parents.

"I wouldn't have told them myself. It just makes Mum worry," Tim told the epilepsy nurse later on the phone. The specialist nurse had called him at his mother's request to see how he was.

He assured the nurse that he had been looking after himself and that the seizure hadn't happened because of anything he had done.

"I'm playing rugby today, or I was supposed to, so I didn't drink last night," he told her. The friend who had telephoned his parents corroborated the story. The seizure was nothing more than bad luck. Seizures are unpredictable. They happen for no reason most of the time. Also changes in lifestyle such as a move to university can sometimes mean that treatment needs to be adjusted. Tim's drug dose was increased. He did not have any seizures for the rest of the year.

Tim passed all his first-year exams – just – and went away for the summer to South America. The trip was not popular with Maureen but Jack and Sean talked her into it. One seizure in two and a half years was hardly a reason to keep him at home like a child, they persuaded her. During the trip Tim had a further seizure. In bed asleep once again – in a dormitory room in Cuzco at the end of the Inca trail. Nobody in the room knew he had epilepsy, but a student nurse who happened to be in a neighboring bunk took care of him. When Tim refused to be taken to a Peruvian hospital she sat with him for several hours after the attack to make sure he was okay. Tim told his brother, Sean, about what had happened. He didn't tell his parents. They only learned about it six months later.

It was a relief to Maureen when Tim came home and went back to start his second year of university. He had moved out of the halls of residence and into a shared house with five friends. Most were new friends he had met in his first year. But one, Jason, was somebody he had known since he was twelve years old. They had gone to school together and played on the same rugby team. They were studying on different courses but still played rugby together at university. Maureen liked having somebody she knew living with Tim. For a while it had made her lonely to think of her son. Picturing him with this old friend made her feel better.

After the boys left school, Maureen hadn't seen Jason for two years. That is why the shaggy top of his head didn't ring any bells when she had first seen it out the window flanked by the policemen.

"Mrs. Dolan?" the police asked again when Maureen didn't answer.

"Mrs. Dolan ..." Jason said at almost the same time.

Maureen registered him properly only when he spoke. His presence meant that she knew why these people were at her door. She didn't know what to say.

"Jason? I didn't recognize you ... your hair's got so long ..."

"Can we come in please, Mrs. Dolan?" one of the policemen asked. "It's about your son Tim."

Maureen could see that there were two cars parked on the road outside the house. Beside the police car was another with a man and woman sitting inside. She knew who they were. They were Jason's parents. She had met them often at school functions and rugby matches and Christmas concerts.

"I think it might be best if we go inside," the policeman said again. He pushed the door, took Maureen by the arm and gently led her into the house. She turned and they all followed her into the kitchen. Once there she turned to face them again.

It didn't occur to her to offer anyone a seat so the two policemen stood face-to-face with Maureen while Jason hung back.

"Is anybody else at home?" the policeman asked and she told them her husband was out. "I am very sorry, Mrs. Dolan, there's no easy way to say this. We are here because your son Tim was found dead in bed this morning. We are very sorry for your loss."

When Tim first left home Maureen had imagined a moment like this. She always thought she would scream or break down in some terribly histrionic way, but when it was actually happening all she felt was numb. First she thought it must be a mistake. She had talked to him on the phone the night before. He had been studying and he told her he was going to watch television and go to bed early.

"Why don't you sit down," one policeman said and pulled out a chair.

Maureen sat. The other officer pulled out a second chair and sat beside her.

"Is there anybody we can call for you?" he asked.

"I should call my husband," Maureen said, leaping up from the chair.

Maureen got her phone. Her hands shook as she tried to put in the pin code to unlock it.

"Maybe I can help?" the officer said.

He took the phone from her and asked what name her husband was listed under. Maureen took the phone back when she heard it start to ring. When Jack answered she couldn't speak a single word. She tried to say Tim's name but started to cry instead. In the end the policeman spoke to Jack, telling him that there had been an accident involving their son and he was needed at home urgently.

The policeman then turned to Jason.

"This is the young man who found Tim, Mrs. Dolan. He wanted to come in case you had any questions. We don't know what happened yet, I'm afraid. We only know that he was found in bed. There will have to be an autopsy."

"He had epilepsy," Maureen said.

It took several weeks for Maureen and Jack to take in the fact that Tim was gone. Because he hadn't lived at home for over a year they could easily tell themselves that he was just away at university and that he would be home at the weekend. But then he didn't come home and the story Jason told her of her son's last few hours went around and around in her head. She found herself searching it for an explanation.

The night before he died Tim and Jason and another housemate had stayed in watching television, just as Tim had told his mother he planned to do. *The Apprentice* was on and the three watched it together. Jason assured Maureen that they weren't drinking and none of them had ever taken drugs.

"Did you see Tim take his epilepsy tablets?" Maureen had asked Jason.

Jason had never seen Tim taking his medication but he had seen the packet on Tim's bedside locker. All three boys had gone to bed shortly after eleven. Tim had seemed in good spirits. He hadn't seemed unwell.

The next morning Tim's bedroom door was closed and he didn't appear in the kitchen. It wasn't unheard of for Tim to miss a lecture so none of the flatmates tried to wake him.

"I think Jason felt very guilty about that," Maureen told me. "He got very tearful when he was trying to tell me."

"He couldn't have known," I said.

"No," she said, "no ... he couldn't."

Jason had only one lecture that day so he went back to the house late morning. Tim's door was still closed. There was nothing particularly odd in that. When it got to midday Jason decided to wake Tim. He knocked on the door. When there was no answer he opened the door and peeped in. He could see immediately that Tim was still in bed. He knocked several times on the door but Tim didn't move. Tim was facing away from him so he walked around the bed calling his name.

"He was lying with his eyes closed. In the dark he looked like he was asleep," Jason told Maureen.

As soon as Jason tried to shake his friend awake he knew that he was dead. His body was already stiff. Jason pulled the curtains open. He saw that Tim's face was waxy and colorless. He phoned an ambulance.

"It's my fault. I wish I'd never let him move away to university," Maureen told me.

"It could have happened anywhere, Maureen, you couldn't be there every minute. It could have happened at home too."

"But I would have checked on him sooner."

"We don't even know what time it happened."

"He would have been better off at home."

"He was twenty. He couldn't stay at home forever. Seizures can be dangerous. Tim knew that but he didn't want to live in their shadow all the time. He loved university."

"Yes, he did."

Sudden unexpected death is a risk that every person with epilepsy faces. It is referred to as SUDEP (sudden unexpected death in epilepsy). On average 1.16 in 1,000 people with epilepsy dies of SUDEP; that's nearly 4,000 people a year in the United States. For those whose seizures are well controlled the risk drops dramatically. They climb dramatically for people who have frequent convulsive seizures, particularly ones that happen at night. The more seizures you have the higher the risk. For some subgroups of people with epilepsy the risk is closer to one in a hundred. The highest-risk group includes those with frequent convulsive seizures who are taking a large number of epilepsy drugs and those who are being investigated for surgery.

The exact cause of SUDEP deaths is uncertain. It is assumed that some are the direct result of a seizure although often there is no proof that a seizure has occurred. It is possible that the brain's regulation of the heart and lungs is at fault. The mechanism that causes death may not be the same in everybody. As the brain exerts control over all our vital organs, brain disease can lead to failure of several vital functions. SUDEP deaths are usually unwitnessed and autopsy does not provide an explanation. An irregularity in the heart rate leading to cardiac arrest is a strong possibility. Or the central control of breathing is affected until breathing slows and stops. Both of these may be preceded by brain activity becoming suppressed, perhaps in the aftermath of a seizure that has gone unnoticed.

There are several case reports of people dying in video telemetry units. Many of those patients have been recorded to have had seizures that were not noticed by staff. After the seizures the brain activity was seen to be suppressed. The brainwaves flattened. That is a common pattern after generalized seizures but recovery is usually quick. In these people it wasn't. Minutes later the heart began to slow and eventually stopped. Cardiac arrest led to death very quickly. Many of those affected are young people. It is one of the many reasons why, even when it feels like no treatment is working, we keep trying new things and never give up.

9

ADRIENNE

Sitting on your shoulders is the most complicated object in the known universe.

—Michio Kaku, physicist

Walter was surprised to find Adrienne out when he got home from work. They were creatures of habit. In the year that they'd lived together they both came home directly from work and cooked dinner together most weeknights. If either of them planned to go out for the evening they usually discussed it ahead of time. But really they hardly ever went out separately. If Adrienne was meeting friends, Walter almost always joined her. It was something Adrienne's workmates teased her about.

"They think we're strange or that you don't trust me on my own," Adrienne had told Walter. "I blame my epilepsy. I tell them you are there in case I have a seizure."

They both laughed at that. It was their shared joke that Adrienne used her diagnosis of epilepsy as an excuse for things. The truth was that at that time, having epilepsy worried Adrienne very little. It barely impacted on her life.

Adrienne's birth was suspected to be the cause of her seizures. The labor had been prolonged and difficult. She came out floppy and blue and had to be resuscitated. The doctors suspected she had inhaled meconium, the first stool a newborn excretes, and had been deprived of oxygen during her birth. Adrienne had been admitted to the special-care baby unit and had spent over a week in an incubator. She was a small and sickly child.

"She came out of me dead," her mother told me when I asked about the birth.

Birth injury is a very common cause of epilepsy later in life. The brain injury it causes can also manifest as physical or mental disability. Her parents had been counseled that her start in life might have consequences. At first they couldn't help looking out for them. Her older sister had crawled when she was six months old. When Adrienne was not crawling at nine months her parents became very worried and took her to their doctor. He said she was healthy. When she began bottom-shuffling a month later and then walked at eleven months they looked back and laughed at their anxiety. In fact Adrienne achieved all of her childhood milestones easily and promptly, but still it took nearly two years for her parents to stop worrying and put her early hospitalization behind them. Adrienne appeared to have survived the trauma of her arrival into the world. She went from being a very sick baby to a strong and resilient child.

The beginning of Adrienne's life had almost been completely forgotten when its likely consequences made themselves known. She was now twenty-two years old. It was New Year's Day. She was at her sister Jeannine's house. Adrienne had complained of feeling unwell. Nobody had understood how unwell she meant until she collapsed. As her sister described it they were both lying on the sofa and chatting when Adrienne just stopped talking. Jeannine said she looked distant.

"She was just sort of staring and blinking and her eyes kept rolling up in her head," she said.

After that Adrienne's head had turned fully to one side, her whole body became stiff and she had a convulsion. An ambulance was called. It arrived in less than ten minutes, by which time Adrienne appeared at least to have woken up. The discussion she had with her sister and the paramedics in the immediate aftermath of the seizure was largely nonsense.

"First she thought we were on a holiday we had taken five years ago," her sister said, "then she thought I was our mum. At one point she went to the corner of the room and pulled down her pajamas and looked like she was going to go to the toilet right there. The paramedics had to stop her and take her to the bathroom. She kept trying to kiss the ambulance man."

Adrienne remembered nothing of this. Once in the ambulance she seemed to start to make sense but even that she couldn't recall later. From her perspective, she woke in the hospital.

As part of the protocol for investigating anybody after their first seizure Adrienne had an emergency CAT scan of her brain. It ruled out the possibility of a brain tumor. She was extensively quizzed about her use of illegal drugs, which she denied.

"Vodka is my drug of choice. Or was," she told me when we met. By that time she had all but stopped drinking alcohol. "I'd had a bit to drink on New Year but not as much as most people. And no drugs. Now I have a glass of champagne at weddings and Christmas and that's about it. I find if I look after myself

I am less likely to have a seizure. So I look after myself. Epilepsy has made me very boring."

A single seizure is not epilepsy. Adrienne didn't get her diagnosis until, about six months later, she had another attack very like the first. She saw a neurologist and went on treatment.

Closer questioning by her neurologist uncovered information that made it clear that Adrienne's convulsions were not the only seizures she had ever had. She had never thought to tell anybody how often she experienced déjà vu. On and off for years she'd had odd feelings of things seeming familiar when they shouldn't be. She thought this was normal. Of course déjà vu is a normal thing that happens to everybody from time to time, so this was a reasonable assumption on Adrienne's part. However, it can also be an occasional symptom of temporal lobe seizures. In Adrienne the subsequent convulsions proved that the déjà vu was an aura, small electrical bursts in the temporal lobe that had not progressed. Only when they spread to the whole brain and produced a convulsion was their importance realized.

Ultimately an MRI scan of Adrienne's brain confirmed that this suspicion was probably correct. Her right hippocampus was bright and shrunken relative to its partner on the left. Adrienne had temporal lobe epilepsy. The two generalized seizures had started focally. The diagnosis was made unequivocally and treatment started. Adrienne did well. The déjà vu spells disappeared and for a whole year she did not have any seizures of any sort.

I met her seven years later. In that time the seizures had been unpredictable. She had years with none and others where she had three attacks. In one of the seizure-free years she lost touch with her regular neurologist. I met her when things were getting worse and she had been referred back to the hospital. By then Adrienne was having more obvious focal seizures. Now the déjà vu was accompanied not only by stomach sensations but also with anxious, fearful feelings and loss of consciousness. Typical limbic system symptoms. She no longer collapsed, but instead wandered in circles, cried and talked incessantly. After the attacks she was often visibly upset and it could take thirty minutes to calm her down.

I feared that Adrienne would be one of the people who never gets better with medication. Mesial temporal sclerosis can be a drug-resistant cause of epilepsy. But you can't say that drugs aren't going to work until you've tried them.

Adrienne also had other problems that her family, more so than Adrienne herself, wanted addressed. There was concern about some of her behavior. She had gone from being happy-go-lucky to somebody prone to erratic moods, angry outbursts and anxiety.

"She also seems a bit paranoid at times," her mother said.

Adrienne looked angry that her mother had told me this.

"She got upset with me because she thought I was talking about her behind her back with Jeannine. She got it in her head that we were planning a holiday together and not telling her."

"One time," Adrienne said, "I got a bit upset until I realized it was nonsense and then I said sorry."

"I haven't even been on holiday with either of the girls since they left uni," her mother said cautiously, and then added, "and it was more than one time."

Clearly Adrienne's epilepsy, once well controlled, was now less so. It was not unreasonable to suspect that both the psychological impact of having the attacks and the physical impact of them on her brain were affecting her personality and her mental well-being. The mesial temporal regions are important to how we process emotions and judge the emotional reactions of others.

"Epilepsy can affect mood and memory and lots of things. Epilepsy is not just seizures," I told Adrienne. "But you've only tried one epilepsy drug and you take it at a low dose, so there's lots of room for improvement."

I advised Adrienne to increase the dose of the medication she was taking and then we waited. If somebody's seizures occur four months apart then you have to wait for as long as eight months before you can feel confident that the change you've suggested has made a difference. For Adrienne things seemed to get better. In a year neither she nor any other family member saw her have a seizure. That was good news. That is not to say that it had solved Adrienne's other problems. She continued to have unpredictable outbursts of anger. Her family feared she was becoming depressed. Or worse than that, she seemed to be imagining things. One week she thought somebody in the family had stolen money from her. That conviction had lasted a full day during which Adrienne tore the house apart looking for the missing money. Another time Adrienne accused her sister of flirting with her ex-boyfriend of three years. Overly suspicious or anxious angry periods like this were not very common for Adrienne but each seemed bigger than the one before.

I arranged for her to meet the neuropsychiatrist and psychologist. Both doctors agreed that her memory was going downhill and they noted how frustrating that was for her. They did not find any immediate psychiatric problem requiring intervention.

I spoke to the psychiatrist about Adrienne.

"I think she's finding her family's interventions a bit oppressive. Her mum is very overprotective. She says that's what's making her so tense. It's hard to know at the moment if that's all it is. She was well when we met but the story certainly smacks of psychotic tendencies. The incident in which she thought somebody had taken money from her bag could be all paranoia. Having said that she managed to justify it to me today and make some sort of rational sense of it."

Eventually Adrienne decided to move out of home, citing her family as the source of the problem. Shortly after that she started going out with Walter, whom she met through a friend.

"When am I supposed to tell a new boyfriend that I have epilepsy?" Adrienne once asked me. "Do I risk scaring them away with the news at the very start? Or do I surprise them when I suddenly collapse and wet myself on a date?"

I didn't know what to advise.

"We're all hiding something when we meet somebody new," I told Adrienne. "A lot of us are hiding things a lot worse than epilepsy. I say tell them as soon as you figure out that they are worth dating for a while."

Adrienne's seizures were generally so well controlled that she decided against telling Walter that she had epilepsy too early in their relationship. He found out about it in bed one morning. They were both asleep when the seizure started. Walter was woken by the sound of Adrienne making a loud bovine cry. He turned to find her lying on her back, convulsing. She was blue around the lips. It had terrified him.

Adrienne woke to find two paramedics standing over her. Walter was nowhere to be seen.

"Do you have epilepsy, love?" were the first words that Adrienne understood. She told the paramedics that she did. They asked her about her seizures and her medication. Seeing that she was lucid they offered her the choice of whether or not she wanted to go to the hospital. She knew there was no point once the seizure was over so she chose to stay where she was.

"Afterwards when I was alone with Walter I wished I had got into that ambulance and got the heck out of there," she told me. "I found myself alone with a very quiet Walter and a set of wee-stained sheets. It was so awkward. In the end I got a taxi to my mother's. I took the sheets with me!"

I have seen many seizures like Adrienne's. They still scare me. I would have been terrified if I was Walter. In a generalized convulsion a person doesn't breathe. Good medical care requires that the person having the seizure be rolled onto their side to prevent them from choking and protect their breathing. But when you try to move a person having a seizure you quickly learn that their body is so rigid and resistant that they seem twice their normal weight. The noises that come with a seizure are distressed. The facial muscles create an unnatural expression that looks pained. Like Munch's scream. Or worse.

Seeing all of that had put a halt to Adrienne and Walter's relationship. He asked to take a break, saying it was too great a responsibility and he didn't think he was strong enough for it. He said she needed somebody better than him. She had assured him that it happened rarely, but his reaction gave her reservations about the relationship too. It had taken a month apart for things to straighten themselves out again. Walter declared that he missed her more than he thought he could tolerate. Adrienne thought about this and refused to take him back at first. It took a week or two for her to come around and admit she missed him too. They started seeing each other again. In the end the whole incident became something that they both agreed had strengthened them.

They were living together when Walter saw her next seizure. The whole experience was very different for both of them this time. Instead of running away Walter went into organized overdrive. He had had a long discussion with the epilepsy nurse specialists about what he should do if this ever happened. When he couldn't roll Adrienne onto her side he turned her head instead. He put a pillow behind her back to support her and timed the seizure. He had his phone ready to call an ambulance if the attack lasted longer than three minutes. It lasted less than one. When Adrienne awoke she was lying in her own bedroom, in the recovery position, Walter watching over her, his telephone gripped in his free hand. The stopwatch was running.

"Fifty-eight seconds" were the first words she registered when she woke up.

Sharing this experience and coming out of it successfully moved Adrienne and Walter's relationship unexpectedly forward. Three months later Walter proposed. Adrienne accepted. The seizures seemed to settle. They began thinking about their future. The wedding plans were in the final stages when epilepsy threatened them once again.

It started the day Walter came home to find Adrienne uncharacteristically absent. Usually she arrived home from work first. He was accustomed to walking in the front door and hearing the sound of the television coming from the kitchen. This evening the house was silent. The front door was not double locked as was usual when they were both out, but when he called Adrienne's name she didn't answer. A quick glance into each room confirmed that she wasn't there. He phoned her but the call went straight to answerphone. Walter was concerned in the first instance but could calm himself by assuming that Adrienne's train was stuck in a tunnel somewhere, or that she had come home and then gone out again to the local shop. He only began to panic when he saw Adrienne's bag hanging on the back of one of the kitchen chairs. Walter looked inside and found her wallet and keys. He checked the house again. He checked the garden. He called a work friend of hers who said that Adrienne had been quite normal at work and had gone home at the usual time. Walter phoned Adrienne's mother who rushed over, followed shortly by her father and sister. They called the police.

Three very tense hours followed, during which the whole family sat in Walter and Adrienne's kitchen trying to impress on two police officers how out of character any such disappearance was for Adrienne. They were all staring hopefully out of the front window when they saw a next-door neighbor come up the path.

"I saw the police car," she said when Walter answered the door, "have you had a break-in? Because I think I have too."

The neighbor explained that she had come home to find her front door was not properly locked.

"I never forget to lock it," she told the police. "And when I went inside a drawer in the kitchen was wide open. I would never ever leave it like that. Never."

"Did you have a look around?" the police asked.

"No, I saw your car outside. And I was too scared."

The two police officers went to the neighbor's house. They called out several times and then began a room search. They discovered a back bedroom door closed. When they tried to open it, it gave a little but seemed to be blocked by something pushed against it on the other side. They heard a whimpering noise. The coincidence that two neighboring houses could be the victims of two unrelated crimes seemed so unlikely that one officer began to call Adrienne's name. The female officer put her identification through a gap in the doorway and beseeched the person on the other side to feel safe and let them in. They pushed harder and the door gave. Adrienne was inside.

The police quickly called Walter to join them and coax Adrienne from where she cowered behind the door. At first Adrienne behaved in a childlike way to the police, seeking comfort and crying. When Walter appeared she became aggressive and struck out, hitting one of the police officers across the face. In the end an ambulance was called and she was restrained and taken to hospital.

In the casualty department Adrienne was found to be part lucid and part paranoid and delusional. She was orientated. She knew who and where she was. She could confirm all the details of her medical history. When asked to explain how she came to be in her neighbor's house, she gave a detailed story. She believed that Walter and the neighbor were having an affair. She had used a spare key that they held for the neighbor to go into the house to search for clues. She was hiding in the spare bedroom to try to catch them together. In the casualty department Adrienne was so upset that she had to be contained in a room on her own. She insisted on being attended only by female staff. When any man tried to approach her she became acutely upset. Walter insisted that there was no truth to Adrienne's story.

"I do not have a clue what she's talking about," he said.

Jenny, the specialist epilepsy nurse, was called and she confirmed that this was all very out of character for the Adrienne she knew. Nor had Adrienne ever expressed any doubts about Walter before. Jenny arranged for Adrienne to have an urgent EEG. Continuous seizure activity in a single part of the brain could leave a person walking and talking, apparently awake, but very confused and unable to function. This is called partial status epilepticus. Adrienne's EEG was utterly normal. She was not in status. Blood tests were taken to screen for drugs or anything that might explain what was going on. Her epilepsy drug levels were checked in case she had taken an accidental overdose. A brain scan looked for infections or strokes. Nothing new or worrying came to light.

A psychiatrist was called to do an assessment. He found Adrienne to be suffering an acute psychotic episode. The features of psychosis are confused, disturbed thoughts, hallucinations, delusions and lack of insight. There are many causes. Often it is seen as a purely psychiatric problem associated with diagnoses like schizophrenia or a bipolar disorder. But each of the features of psychosis can also happen in people as a consequence of physical disease or other provoking factors. A healthy person who is sleep-deprived or exposed to extreme stress or drugs could hallucinate or become paranoid. Medical problems such as thyroid disorders, autoimmune conditions, tumors, strokes and a variety of hormonal imbalances (pregnancy amongst them) have also been implicated.

Adrienne had a delusional conviction that Walter was cheating on her. She was also suspicious of all men, believing there was some conspiracy at hand that she couldn't understand. Her train of thought was working at high speed and she jumped from subject to subject. Her mood was very low and several times during the conversation with the psychiatrist she threatened to hurt herself. She did not want to go home. If they sent her home she would kill herself. The psychiatrist judged that her threats were empty and that the real risk was low but still she should be kept in hospital for observation and further investigation. Adrienne was started on an antipsychotic drug. She spent a week in the psychiatry ward. She was lucid again within five days.

"Do you think she's been having seizures?" the psychiatrist asked.

"She says not, and Walter hasn't seen anything since the last convulsion over a year ago," I answered.

In somebody with a brain disease it is illogical to try to extricate a psychiatric problem as something entirely new and separate. I needed to figure out what relationship Adrienne's psychosis had to her seizures.

"You've had no seizures for a year?" I checked with Adrienne when she was well enough to communicate with me again. She remembered much of what had happened in the previous week, but not what had driven it.

"None," she said. "Sometimes I think I feel the déjà vu, I suppose, but I'm never sure. I have had no definite seizures."

"She's been sleeping badly," Walter told me. "She sits up in bed at night and mutters and lies down again. She doesn't always remember it."

Once Adrienne felt better she had apologized to Walter for her accusations and they were quickly back to normal.

"I don't know why you thought that about me," he said.

"I don't know either," she told him.

It was my job to answer that question. I suggested that I arrange for some further EEG recording. Nobody on the psychiatry ward had seen her have a seizure, but perhaps there were some before she came into hospital. Many patients with epilepsy have seizures they cannot detect in themselves.

As usual in medicine the answer depended on giving the story time to unfold. Adrienne came into the video telemetry unit. Her paranoia was behind her. I knew I might be intervening too late. She was now on a low dose of an antipsychotic medication along with her usual epilepsy drug. The morning after her second night of monitoring the technician came and told me that Adrienne had had a seizure.

"Did she know?"

"I don't think so. I asked her how the night went and she didn't say anything about it."

"Okay. Let's keep recording and see if this is a more common occurrence than she realizes. Don't tell her if she has a seizure. I want to know what she knows about it."

Unlike other illnesses, the manifestations of epilepsy can be invisible to the person who has them. The brain is the organ of consciousness. It has to be awake in order for us to give an account of ourselves. For Tina, another of my patients, that was the conundrum. She had been in the video telemetry unit shortly before Adrienne. During her week with us she pressed the event marker thirty times to alert the nurses to the fact that she was certain she had just had a seizure. On none of the thirty occasions could we find evidence either in the video or the brainwaves that anything untoward had happened. Sometimes she was reading or watching television when she would suddenly look for the alarm and press it. More than once she fell asleep in her chair and woke with a start and pressed the alarm button. On each occasion Tina reported that she had been unconscious for at least a minute but there was never anything to see on the video to support that claim. She was imagining it.

That is not to say that Tina had no epileptic seizures. She did. Lots. But she had been oblivious to those. They were reported to me by the staff. Tina did not press the alarm for any of them. To an onlooker they were overt. She chewed vigorously. One arm rose slowly into the air and then began to jerk. She made an unnatural glugging sound. They weren't apparent to Tina. I had to tell her about them. All the time at home and in the street, people tell her that something has happened. She tries to detect it in herself. Unable to spot it, she begins to guess. She nearly missed her bus stop – was that a seizure? She can't remember if she turned the heating off – was that one? It is simply impossible for her to know.

"I almost wish nobody told me when it happened," Tina once said. "It's like having somebody tell you that you've walked around all day with food stuck in your teeth. If you never found out there'd be nothing to be embarrassed about."

Adrienne's seizures had never been hard to miss. If she didn't collapse and didn't have déjà vu then she assumed she didn't have a seizure. It seemed as simple as that. The second night of her stay in the video telemetry unit had already proven her wrong. She had an attack of which she was unaware. The third night none. The fourth night none. The fifth she had a cluster of six seizures in six hours. The EEG discharge arose in the right temporal region each time.

Each attack was very similar. Adrienne was asleep. She woke and opened her eyes. As the electrical discharge evolved through her right temporal lobe and then spread to most of the right hemisphere of her brain she did very little except blink and occasionally stiffen very subtly under the bedsheets. There was no way that this subtle movement would wake a sleeping partner. Even if Walter did wake there was nothing very obvious to see.

After the fifth seizure Adrienne seemed to wake more completely. She sat up and appeared to be looking for a light switch. She spent a long time pressing anything she could find. The nurse call button. The video telemetry alarm. She played with her mobile phone for a while. She knocked cards on to the floor from the bedside locker. After a while she climbed out of bed and became tangled in the wire attaching her to the equipment. A nurse responding to the alarm came and asked her if she was okay. It was no more obvious to the nurse than to Adrienne herself that she had just had a seizure. The nurse told me the next morning that she assumed Adrienne was just a bit confused as a result of waking in a strange place.

After the sixth seizure Adrienne walked to the door. She seemed to get upset. The fact that she was tied to the room by a wire seemed to frustrate her. She disappeared out through the door and came back again a few times as if trying to solve the conundrum. A nurse tried to bring her back to her bed but this time she resisted. She insisted that she needed to go to the toilet. She couldn't understand the nurse's explanation that her bathroom was en suite. Several times she called one of the nurses Jeannine. When she was corrected she righted herself. "I know you're not Jeannine. Jeannine is my sister." After a while she went back to bed and fell asleep.

The following day I told Adrienne about the seizures and suggested that we needed to change her treatment. She was surprised. Disappointed. She hadn't even suspected it.

"I really thought they were gone."

"Better we know, now we can do something about them."

I scheduled for Adrienne to go home the next morning but her bad luck was not yet quite over. Overnight she had become inconsolable. Unreachable. Making very little sense. She was even aggressive. When a nurse approached her she had struck out.

"Another patient visited her room. They got in a fight. Adrienne got some idea in her head about the other patient. Not sure what. I saw them arguing on the camera and when I went into the room to find out what was going on Adrienne took a swing at me."

"Oh no."

"Oh no is right."

"You okay?"

"Oh yeah, I'm okay. She didn't get me. But we can't handle her. She's very upset."

I went to see Adrienne in her room. She launched herself at me in relief. She was agitated. She started pacing the room and wringing her hands. She told me a long story about a missing wallet. I could see a wallet lying on her locker.

"Is that the wallet?" I asked.

"Yes, I got it back."

"Are you sure somebody tried to take it?"

Adrienne walked to the window and back. Instead of answering she recounted the story of the wallet again as if I hadn't just heard it.

"That nurse is a cow. She wouldn't help."

Adrienne began writing in a notepad. I could see that the page was already almost completely filled with tiny writing. I glanced at what it said. It was too small to read properly but I gathered it was a further account of her disagreement with the nurse.

"You can take this to the person in charge," she said, tearing a page from the notepad and giving it to me.

I talked to her but couldn't get her past the point of accusation. I went back to discuss the situation with the nursing staff.

Jenny came to sit with Adrienne and helped quieten things down by promising a full investigation. Of course no detailed investigation was needed. Everything that happened in the room had been recorded. The technician had already told me that she could find no justification for any of Adrienne's accusations.

I looked through the video for myself. From early on in the day Adrienne seemed out of sorts. Little things made her tearful. Dropping a slice of bread at breakfast time. A phone call. Before the other patient had paid her a visit she already seemed volatile. The patient had seemed placating rather than argumentative. When things had blown up they had done so very suddenly. I could not make out every word that was said, but I could see the rising tension in the body language. When the nurse came in to see what was going on between the women, Adrienne was shouting while the other woman had her hands raised, seemingly in an attempt at conciliation. When the nurse approached Adrienne, things quickly escalated. Whatever the nurse said made both of Adrienne's hands rise into the air, one pointing to the door and the other pushing the nurse roughly backwards by the shoulder. As I watched the video I also watched Adrienne's brainwaves. In contrast to the previous night of seizures they were entirely blameless. The psychiatrist was called urgently.

"It's a postictal psychosis," the psychiatrist confirmed.

Ictal is Greek for "a blow." It is the word we use to indicate when the seizure strikes. The seizures were gone but they had disrupted her brain chemistry in such a way as to leave Adrienne irrational and changed.

Ictal psychosis occurs during the seizure – where abnormal brainwaves stimulate the brain to produce hallucinations or irrational behavior. Postictal psychosis is something that happens to the brains of some people a day or so after they have had a seizure. They appear to recover from the attack but its effects linger unseen. It is commonest in temporal lobe epilepsy, particularly where there has been a cluster of seizures. Because there is a lucid period of twenty-four to forty-eight hours after the seizure, the psychosis, when it happens, appears to come from nowhere. It is an ill-understood consequence of an electrical surge through the brain. Some believe it is related to a transient derangement or depletion of neurotransmitters caused by the seizures.

Gamma-aminobutyric acid (GABA) is the principle inhibitory neurotransmitter in the brain. Both epilepsy and psychosis are related to an imbalance between excitation and inhibition of brain cells. Drugs that promote GABA in the brain are used to treat epilepsy. They make it harder for neurons to produce the synchronous electrical burst that leads to seizures. A deficit of GABA may also have a role in causing postictal psychosis – and, for that matter, all types of psychosis. Abnormalities in cortical inhibition have been described in schizophrenia and postictal psychosis. The exact mechanism by which a seizure makes a person psychotic is unknown, and attempts to explain it are all speculation. Other possible causes include changes in cerebral blood flow, or a hypersensitivity to dopamine, the neurotransmitter we rely on for our motivational and reward system.

Postictal psychosis does have features that distinguish it from disorders like schizophrenia. It behaves as seizures do. It is abrupt in onset, short-lived and self-terminating. Popular culture likes to link violence and seizures, but ictal violence is very rare. Aggression exhibited during a seizure is almost never purposeful or directed. Somebody having a seizure might kick out at another person, but they could just as easily have kicked a chair or the wall or the air. Postictal psychosis, however, has been associated with directed violent behavior. Driven by persecutory delusions and misled by hallucinations, a patient in postictal psychosis may appear to act with deliberate intent. But while actions may seem to have purpose they are entirely without insight or meaning. They are random.

Adrienne was given an additional dose of antipsychotic medication. Jenny helped to calm her down and arranged for Adrienne to have one-to-one care from then on. Staff would sit with her overnight.

Her parents and sister visited. That helped. But when Walter arrived she refused to have him in her room. Overnight Adrienne didn't sleep. Every now and then she woke and became upset. She thought she had heard the nurses and other patients talking about her. She no longer wanted the one-to-one carer in her room. It was distressing to watch. Most of the time that she was awake she was scribbling detailed accounts of everything done and said in her notebook. She had nearly filled it when the additional dose of antipsychotic suggested by the psychiatrist seemed to take effect. The last psychotic episode lasted five days. This one wore itself out in two.

We know that our thoughts and consciousness, our personalities and occupations, are inextricable from the physical reality of our brains. But, even knowing that, it sometimes feels that we can't take the next step to what that really means. If you have a brain disease you are very vulnerable to developing a mental disorder alongside it.

People with brain diseases commonly develop psychiatric symptoms. In multiple sclerosis, mood can be either euphoric, abnormally high and disinhibited, or low. Anxiety, agitation and severe depression are much more common amongst people with MS than they are in the general population. Those affected by Parkinson's disease are prone to mood problems and psychosis. Attempts to treat the disorder can also produce issues with impulse control. This manifests as gambling or compulsive shopping, binge eating and hyper-sexuality. Some Parkinson's sufferers are affected by punding – an intense fascination with disassembling and reassembling equipment, collecting seemingly banal objects, hoarding and the compulsive sorting through of belongings. Lots of neuro-logical conditions present as apparent purely psychiatric problems in the first instance – dementia, Creutzfeldt–Jakob (mad cow) disease, autoimmune encephalitis.

The brain is the organ of the mind. The mind is made up of many elements: thinking, judgment, intelligence, memory, emotion, reasoning, perception. The fondness for tracking brain function with fMRI has also been applied to these elements of brain function. Each can be given at least a notional location in the brain. I think it makes us feel better to have a biological mechanism for everything we do. And yet, despite the inextri-cable nature of mind and body, many health services treat the two as if they were unrelated. Physical and psychological medicine can be kept so separate that they are even based in geographically separate institutions.

People with epilepsy have many psychiatric co-morbidities. The brain does not come out unscathed from the assault of the electrical discharge. Over a hundred years ago Hughlings Jackson noted that "acute attacks of insanity" could be provoked by seizures. It was once called epileptic fury. In video telemetry units where drugs are withdrawn in a deliberate attempt to provoke seizures, the resulting seizures have sometimes been followed by hallucinations and delusions and anger. Not as a feature of the seizure itself, but an aftermath.

The treatment of psychosis in epilepsy is first and foremost to prevent the seizures. If the seizures cannot be controlled then the psychosis is treated with the same sorts of drugs as in any other cause of psychosis. In the eighteen months that followed, Adrienne had two more psychotic episodes. Her background level of anxiety also seemed to increase. She became filled with hyperactive anxiety. When I met her in clinic she was a ball of energy, trying to look on the bright side, but clearly struggling.

"Walter called off the wedding," she told me at one visit. "He said that nobody could live with me, I'm too difficult."

She smiled as she said this. Laughed even. I was so upset for her that I had to make an excuse to leave the room and call one of the epilepsy nurses to join me. The nurse knew Adrienne well, so after we had talked together for a while I left them alone. The nurse told me later that Adrienne said Walter had been unable to handle the unpredictability, and blamed the break-up on her.

Adrienne had to move back to her family home. That caused her to slide into depression for a while.

"My life is going in reverse," she told me.

Living with her parents made her feel like a child. I had encouraged the move. Pushed for it, even. So had the epilepsy nurses. If Walter was not there to notice when Adrienne wasn't well, what would happen? Maybe nothing. Somebody else would find her. Someone else would notice she was missing and look for her. She would probably be fine. But maybe she wouldn't.

Three years after Walter's abrupt departure the surgeons stepped in to save the day for Adrienne. The shrunken hippocampus made her a good surgical candidate. Adrienne wanted her independence back. She wanted to go out without having her mother texting to check up on her. She wanted to live alone. She wanted to drive a car. She wanted to take off her medic-alert bracelet. Surgery that offered a seventy percent chance of seizure freedom felt very much worth it to her. It worked. Adrienne's seizures and her psychosis were successfully cured by the removal of a small section of her temporal lobe. And she hasn't missed any part of that absent piece of brain.

10

MIKE

If a man has lost a leg or an eye, he knows he has lost a leg or
an eye; but if he has lost a self – himself – he cannot know it,
because he is no longer there to know it.

—Oliver Sacks, *The Man Who Mistook
His Wife for a Hat* (1985)

Mike's brothers and parents were by his bedside when he woke
from a weeklong coma spent in the intensive care unit. There
was widespread relief. They made phone calls to the rest of the
family to tell them the good news. They called his girlfriend
of six months and asked her to join them.

"He doesn't know what happened but he is able to talk and
he asked for a drink of water, and he knows who we are and
the doctors think he's going to be okay," they told her.

It was a celebratory atmosphere in the intensive care unit that
day. And the celebration continued. A week later, against all
that had been predicted, Mike was walking around the general
surgery ward. He was conversing normally. In fact Mike was
asking to go home. He felt ready. His doctors were against it.
The surgeon who operated on him felt that he should transfer
to a rehabilitation ward and go home from there. He would

have to wait on the surgical ward for a place to become available. Nobody could say how long that would take. Mike was insistent that it wasn't necessary and that his recovery would be quicker at home. He was convinced that what he needed was normality. He had survived something that people said he would never survive and he wanted to get out of the hospital and enjoy being alive. The doctors remained reluctant. The situation resolved itself when Mike signed his own discharge papers against medical advice. He went home at his own risk. Of his three weeks in hospital Mike had spent one in a deep coma, supported by machines, with doctors keeping him alive. As he walked out the front entrance of the hospital and climbed into his father's car the only evidence of his accident was the shaven right side of his head and a discreet bandage, both of which were hidden by a baseball hat. Mike and his family believed that was the happy ending to a terrible story. It was indeed an ending. But not in the way that the family thought.

Mike was a lawyer. He came from a reasonably well-off, middle-class background. He was the youngest of three brothers. The older two were a year apart. Mike came six years later. He had the confidence that comes with being the much younger child. Born after his brothers had already started school, Mike got the full attention of his mother. She was a nurse but did not return to work until all three children were in secondary school. His father was also a lawyer. Both parents valued education and pushed their sons to succeed. Mike was naturally bright. At every opportunity he did well.

After studying law Mike had taken a high-powered job in the financial center of London. The job required that he work long hours in a competitive environment. He was still in his twenties and fed off the adrenaline. At key times of the year it was not unusual for him to stay at work until long after midnight and then return to the office at seven the next morning. In the evenings cars were sent from the office to top London restaurants to bring dinner to any workers staying overnight. Work parties were lavish, as was remuneration. These were the rewards given to those willing to give everything to their job.

Mike lived that life for ten years. By the time of his head injury he was nearing the top of his profession. His life was full of financial reward and security but it was not easy. He made many personal sacrifices for his career. It threatened relationships and took up all of his time. He was under constant scrutiny and pressure. Mike was not naïve. He knew his were First World problems. He was luckier than most people. But that did not mean that he was invincible.

One morning in December a few years ago Mike got up to go for a run. It was a Saturday. The streets were littered with the detritus of the previous night. It was early and Mike was out amongst the service workers and stragglers returning home from the night before. His usual run took him on a route from his apartment, around several blocks of his London neighborhood, through a park and home again.

Nobody knows the full story of what happened to Mike that morning. All that is known is that he did not complete his run. Instead he was found collapsed in a doorway fifteen minutes from his home. A street cleaner discovered him, thinking that Mike was either drunk or had taken drugs. The cleaner was concerned because Mike was unrousable on a freezing morning, so he called an ambulance. He then waited with Mike until the ambulance arrived. The paramedics were the ones who noticed the laceration to Mike's head. They could not get a coherent response from him so they rushed him to the nearest major accident and emergency department.

Once in the hospital Mike was fast-tracked through the system to have an urgent CAT scan of his brain. It showed a collection of blood inside his skull. The accumulation of blood was putting pressure on his brain so that the right hemisphere was distorted and pushing over the midline, compressing the left hemisphere. Mike needed urgent life-saving surgery to remove the blood clot. The operation relieved the pressure but there was no telling how badly the brain was damaged or if Mike would ever wake up.

During all of this – the scans, the surgery – nobody knew who Mike was. The name over his bed was John Doe. When he was found he had no identifying information, no wallet, no possessions. Medical staff briefly speculated that he was home-less, but his general health and his good-quality running shoes belied that. A drug and alcohol screen showed no evidence that he had taken either that morning or the night before. Nobody knew what had happened to him and the police had no way of identifying him. After surgery he was transferred to the intensive care unit where he waited for somebody to miss him.

It was probably about the time that Mike was going into the scanner that his girlfriend, Zoe, woke up and called out for him. They had been out together the previous night and she was staying at Mike's apartment. Mike had got out of bed before she was fully awake. He had whispered to her that he was going for a run but she had only half-heartedly woken to acknowledge it. When she woke properly she didn't know how long he had been gone. She waited an hour and then put on some clothes and wandered out into the street to see if she could find him. She had absolutely no idea what route he had planned to take. She meandered for fifteen minutes before realizing how fruitless that was. She rang his phone but it was unavailable. She was in a quandary after that. She didn't know his parents well enough to feel comfortable contacting them. Anyway, she felt stupid. What if he had simply met somebody he knew and stopped for a coffee?

An hour later Zoe felt certain that something was wrong. She rang a friend of Mike's. They gathered a search party in the local area. Soon they started ringing the local hospitals. The very first hospital they rang, the closest, had admitted a John Doe matching Mike's description. Mike was already out of surgery and in the intensive care unit when his horrified family came to identify him. He was John Doe for half a day before the nurses were able to rub out this tentative pseudonym on the board over his bed and give Mike his proper identity. They didn't realize that Mike's identity had already fundamentally changed beyond repair.

*

I met Mike a year later. In fact we had two failed meetings before he finally walked into my office. The first time he came to my clinic I was running fifteen minutes late. A previous patient had run over time and it had a knock-on effect. Most patients don't mind waiting if staff are polite and apologetic and, most importantly, when they are seen they are given a fair share of time. Mike had not been willing to accept the delay. An apology did not placate him. He wouldn't tolerate the inconvenience. An ugly argument broke out between him and the reception staff. I could hear a commotion from my room but since I was with another patient I couldn't look out to see what it was. When I was finally ready to see Mike I was told he had left.

"He said he couldn't wait. He was pretty forceful. He said he was going to make a complaint to the chief exec," the care assistant who looks after my clinic told me. I love the care assistant who manages my clinic. She is endlessly efficient and good with patients. That Mike had defied her calming abilities said a lot.

"So be it," I sighed and moved on to the next person.

A few weeks later I received a letter from his GP. They had heard that the appointment with Mike was unsuccessful, and they would be grateful if I could arrange a further appointment. They explained that Mike was struggling since his head injury and asked for my patience. I sent another appointment but that second attempt at meeting was as disastrous as the first. This time Mike arrived on time, the clinic was running promptly, but for some unexplained reason he left and never reappeared. I never found out where he had gone.

On Mike's third clinic appointment we finally met. This time he came with his parents.

"Don't ask me why these two old people keep following me around," he said, laughing, when I invited them into the room.

"Is it okay that we're here?" his mother asked, looking abashed and worn out.

Mike turned to her.

"I've told you that it's patently not all right with me – but here you are. If a doctor tells you that I'm an adult and able to account for myself will you believe her over me?" Mike said.

I was immediately in a dilemma. He was making it clear that his parents were unwelcome. Was I right to include them? I wanted them there. At the very least they could fill in the blanks of Mike's missing week spent in intensive care. And much more, too, I suspected. I tried a cautious question.

"Perhaps your parents could stay just to let me know the details of the time you spent in intensive care? There may be things you can't remember."

"You're all in cahoots," Mike was laughing, but it wasn't a very good-humored laugh. Then he abruptly turned and pointed his finger so that it was an inch from his mother's face and then his father's. "I'm permitting you to be here. I am. Me. But I do the talking. You are not in charge here."

We agreed on this compromise. His parents would remain in the room but the consultation would be between Mike and myself. Throughout the negotiation his mother looked like she might cry at any moment. Mike meanwhile sat in his chair and rocked backwards as if he were asserting his place in a boardroom. He looked the part for that. He was dressed quite formally for the setting. He was young and handsome. Being authoritative sat very easily with him. His hair, shaved for his surgery, had grown back a long time before and I could see no external sign of his head injury. Nor was there any particular immediate clue to how disabled he was. In fact he did not look disabled at all. Anywhere but in that room he would cut an impressive, even enviable, figure. But looks can be deceptive.

After Mike had gone home from the surgical ward he and his family had initially celebrated how unscathed he seemed.

"When we saw him in intensive care we were told to expect the worst. We called everybody who mattered to come and say goodbye to him. That's how bad things were," his mother told me, looking cautiously at her son as she spoke. He allowed her to contribute this. It spoke to what he had been through and what he had overcome.

"They absolutely expected me to die. It's a miracle I'm alive. I'm a miracle!" Mike looked very pleased with himself as he said this.

Mike felt so well that, if he had had his way, he would have returned to work a week after coming home. Zoe and his family persuaded him that he needed more time. In the end he took sick leave for a further month and spent that time trying to find out what had happened to him.

There were no witnesses. CCTV footage in the area showed Mike running down a high street near his home. A second camera caught him on another street and then the cameras lost him. There were other people in the area but Mike did not interact with any of them. Police could only conclude that he had been mugged. Mike had been listening to music on his phone as he ran. Pictures showed it in his hand. It was missing when he was found and it never turned up. It was possible he just tripped on an uneven paving stone and that an opportunist took his phone, but there had been two other muggings in the area that month, so that seemed most likely. The police guessed that he had been pushed and fell, hitting his head on the concrete. Whichever scenario was correct the force of the head injury was enough to cause a large intracranial hemorrhage that rendered Mike unconscious.

Seven weeks after the accident Mike returned to work. Cake and balloons were waiting for him. "We knew you were a hard nut to crack but now we have proof" the welcome-back card said. Mike's return was part-time at first. That was not his choice. He had been keen to maintain an air of invincibility but was foiled by protocol. Mike planned to prove himself over the next two weeks and build back to the usual workload as quickly as possible. In fact he barely lasted two weeks before he was asked to leave and not to come back until he was fully recovered. He left without saying goodbye, not realizing that he was never going back.

It had been difficult for Mike's colleagues to know how much leeway to give him in his first few days. He was a man recuperating from a physically and psychologically traumatic incident, so why should he not struggle at first? Mike was given the lightest tasks but couldn't seem to follow them to their conclusion. Occasionally he finished but more often he was distracted by something else and became absorbed in that for a while before ultimately moving on from that too. His colleagues began supervising him, much to Mike's chagrin. It was difficult to criticize him since he was as enthusiastic as a young child in everything he did. Finally Mike made a mistake that was sufficiently remiss to threaten both a business deal and the integrity of the company. Mike was asked to take more time off. His reaction to this suggestion proved problematic. He argued his position to the point of becoming heated. He presented circular arguments and endless justification for what he had done and why he should stay. A friend had to escort him to a taxi and send him home.

At home Mike seemed to lose all his motivation. He had already lost interest in playing sport and running, but now he didn't want to go out either. He took to watching television most of the day. His family wanted him to go back to see the neurosurgeon but Mike insisted that he didn't need to see any more doctors. Friends who visited waxed endlessly about how well he looked. Zoe was frustrated and was having difficulty tolerating his newly unpredictable temperament. One minute he was affectionate, but the next he got angry about something small. One day they were watching a movie together and he started to cry. Mike had been a person of steady moods who rarely shared any great outward displays of emotion. He had never been easy to anger and, in their time together, she had never seen him cry. She concluded that he was depressed.

Mike had never been idle in his life. She was worried it was taking its toll. She contacted Mike's brother and asked if the family would talk to him about seeing a psychiatrist, or even just his GP. The family were reluctant to intervene too actively. Instead Mike's brothers took to visiting more often and forced him to go to the park for games of touch rugby. Spending more time with Mike made them share Zoe's concern. They noticed that he was more argumentative. They put their trust in time to heal that.

There was one small plus. Everybody agreed that, on a good day, Mike was more fun than he had been. He had always been driven and ambitious above anything else. Now he had a jovial side. He was quicker to get angry but he was also quicker to burst into a belly laugh.

Four months after the accident, despite his brothers' intervention, Mike still wasn't well enough to return to work. He'd also lost the will to go back. He even stopped contacting his workplace to discuss it with them. His nascent relationship with Zoe didn't survive. After a furious argument over a trivial matter she asked for a break.

It was a week or so after that, when he went to his parents' house for Sunday lunch, that they were shocked to see a large bruise on his forehead. He also had what looked like a carpet burn on his arm.

"What happened?" his mother asked.

Mike shrugged off the injuries, so the family did too.

A month after that he complained to his mother that his tongue was painful. He stuck it out and she was shocked to see deep purple tooth marks down one side. With Mike still unwilling to see any doctor she went to her own GP to discuss her son's problem in principle. Her GP had never met Mike but agreed that she was right to be concerned. At their parents' urging, Mike's two brothers wrangled with him and got him to his doctor. They were fortunate to get him on a pliable day. When he was pushed into his doctor's office he became quite loquacious, which surprised everybody.

After hearing the story the GP said that seizures were a potential consequence of serious head injury and that Mike should see a neurologist. After those two failed attempts to meet, Mike and I finally did. The appointment was not a success. Before I had made any significant progress Mike diverted the conversation. He began by telling me about his GP, a doctor I had never met.

"Next time you see Dr. Jenkins just say the words 'big bang' to her and I guarantee you that she'll crack right up," Mike said to me after a very brief preamble.

I didn't know what he was talking about. I didn't ask since it didn't seem relevant.

"Do you remember how you got the bruise on your face, Mike?" I asked, trying to get the conversation back to what was important.

Mike turned to his parents, "Dr. Jenkins gets me. We can have a laugh together. She's a good doctor – for a woman!" and he burst out laughing and looked at me again. "I'm joking, I'm joking! But honestly just say the words 'big bang' to Jenkins and she will know what you're talking about. You don't even have to say my name."

"Do you know what your family are talking about when they say that you've injured yourself without realizing it?" I asked.

Mike answered as if I had not even spoken.

"Just say my name, just say Mike P. says hi and see what she says," he said, stuck on a single track.

"Mike, the doctor is trying to ask you about the burn you had on your arm," his father said.

"Sssh," Mike said to his father. "This is between me and this doctor."

"He's not always like this, it's just because we're here," his mother interjected. "And we had a bit of an argument about a parking space on the way here and that seems to have upset things. He had been okay before that."

"We should have parked where I said! And anyway, shh you two." Mike turned. "What was our agreement?!"

Things continued like this. Tiny fragments of information followed by stories that seemed to have very little to do with our discussion.

"Mike, can you appreciate what your family mean when they say that you have not been yourself since the assault?"

"Let me tell you about these two," Mike indicated his parents with the cock of a thumb. "My father has been a solicitor most of his life. He likes to be right. He spends every day arguing that he is. His good wife – my mother – worships him and also likes to believe he is always right."

Oddly this comment broke the atmosphere for a moment. Both parents smiled as if they recognized a little of what Mike had said.

"What is it they think they are right about?" I asked him.

Mike seemed unable to answer. He immediately went into another unhelpful anecdote. Time had long ago run out and I was making no progress. I needed to come to a conclusion.

"Head injuries as severe as the one you've described have more than a sixty percent chance of causing epilepsy. Nobody has seen you have a seizure but the tongue-biting you describe is very typical of a seizure, so I think on balance I should start you on a drug for epilepsy. Seizures are dangerous, especially since you live alone, so I need to try to stop them if they are happening. I also want to do some tests. Is that okay with you?"

From the corner of my eye I thought I saw his mother nod. Sometimes in difficult consultations you find yourself complicit with relatives when you have no right to be.

"You think I have epilepsy?"

"Your head injury was so severe that I think it has caused you to develop seizures. I want to do some tests and start you on treatment for epilepsy."

"Children get epilepsy."

"Adults do too."

"I'm not sure I want to take medication just on your say-so."

"If you are having seizures and you collapse at home when you are on your own it could be very dangerous. What I'm suggesting is something that will help you to keep your independence."

Mike agreed. I suspected that any suggestion that offered a chance to get back to his old life would be well received. It must have been very difficult for him. I couldn't imagine how it would be if one day I was working and the next day illness or injury put me out to pasture.

"I would strongly encourage you to go to the head-injury service too. I think they could do wonders to help you through some of the difficulties you have had since the operation."

Mike burst out laughing, "How many doctors do I need? Aren't you my head doctor now? I only have one head, Doc!"

Six months before Mike had been responsible for millions of pounds' worth of business. Now I was struggling to convince him of simple concepts around his health.

"Maybe one step at a time," Mike's father suggested, tentatively. Mike let this comment pass.

"Okay, let's agree that I will start you on a drug for epilepsy on a balance of probabilities. I can stop it if I'm wrong. I will also get hold of your old brain scans and arrange some further tests. How does that sound?"

"That sounds like a plan."

I went over some instructions. Mike left with a prescription in his hand, looking happy. The door had no sooner closed behind him than it opened again and his mother slipped into the room.

"It really is impossible ..." she started.

"I know it is. I promise you I can see that and I will do whatever I can, but for now—"

Mike burst back into the room. He pointed his finger at his mother. "Are you attempting to tamper with the jury, Mother?"

He ousted his mother from the room and closed the door conclusively.

I arranged for a series of tests. While I was waiting for them to be done I received Mike's brain scan results from his original hospital. It showed extensive contusions (bruising) of his left temporal lobe and both frontal lobes. Broken bones, torn skin, these things heal. The scars they leave are largely asymptomatic. Not so the brain. It heals badly and incompletely, and scars matter.

Mike came to the hospital for follow-up investigations. His MRI confirmed that he had extensive brain damage. His EEG showed changes reflecting that, but no clear evidence of epilepsy. Mike's father emailed me to let me know that Mike had reported waking up on the floor of his flat with no idea how he got there. Mike had become unreliable so the family were not sure if this was true or not. Sometimes he deliberately stoked their concern. I discussed the problem with the psychologists who had also met with Mike. I needed advice. So far I had assumed that Mike had the capacity to make decisions for himself, but maybe he didn't.

"He has capacity. But you need to make sure that all the issues are explained very carefully and precisely. He doesn't register everything the first time. I think he needs to be spoken to in a quiet room without distractions."

I decided a meeting with a social worker would also be appropriate. Mike and his family balked at the suggestion. The words "social worker" meant something different to them than they meant to me. To my mind a social worker could suggest innovations that could keep Mike safe within the life that he knew. In a flat he owned, living alone, independent from family. They might even have some good ideas to help him get back to work. In neurology, where there are so few cures, sometimes the social worker is more valuable than the doctor. Mike could not see it that way. I put that plan on hold.

When Mike's next clinic appointment finally came around I was upset when he didn't turn up. My secretary rings all of the patients a few days before the clinic to make sure everybody remembers they are expected. Mike had confirmed that he would attend.

"There's a message about him on the machine," my secretary told me when I returned to her office after clinic. "He's been arrested. I couldn't really understand the whole message. It was his mum and she sounded really upset."

I rang the number she had left and spoke to Mike's father. Mike had been accused of attempting to sexually assault a woman in the park.

"He just frightened her. He said he just wanted to talk with her and he made some sort of inappropriate remark and couldn't seem to understand when she didn't find it funny. He's a big guy, you know, you can see why a young woman would be intimidated by him. As far as we can work out he wouldn't back off when she told him to go away. I think he grabbed at her. But he didn't mean anything. I know he didn't. This would never have happened before the mugging."

"Of course it wouldn't," I assured him. I had a copy of the neuropsychologist's report in front of me. *Severely impaired frontal lobe function.* "I am quite sure this is a direct result of how badly his brain has been damaged. The frontal lobes are vital in controlling behavior and helping us respond appropriately to others. This is a symptom of his head injury and we need to make sure that everybody understands that so he gets help, not punishment."

"He can't go to jail. Or get an arrest record for sexual assault."

"Let's just make sure that everybody concerned knows the impact of his brain injury and take it from there. I agree, a criminal record or being sent to jail would do him more harm than good."

Disability is not only physical. A person might be very severely disabled by brain disease without having any physical impairment at all. Invisible disability can be very difficult for people to appreciate.

Mike had severe damage to his frontal lobes, the control centers for our executive functions. These include our ability to plan, to judge social situations, to multitask, to be socially appropriate. We all have the occasional naughty impulse and it is our frontal lobes that stop us carrying them out. The frontal lobes help us to learn and keep to rules. They also regulate sexual impulses. A person can retain their intelligence but lose their judgment. Mike remained eloquent in conversation. Deceptively so. It fooled people. Mike looked far too well for how unwell he really was.

The invisible disabilities produced by frontal lobe damage can make life very difficult and dangerous. Judgment can be very weak. At home alone cooking dinner Mike might not be able to tell when food needed to be taken out of the oven or when a pot was too hot to pick up. Planning to do two things at the same time, fill the bath and put the dinner on, could stretch his planning capabilities so that one or other task would lead to disaster. Nor could he rely upon himself to respond quickly or appropriately if disaster did occur. Outside the home, especially with those he did not know well, social situations were likely to be very hard for him to negotiate. Mike would not necessarily have any idea that he was making another person feel uncomfortable.

In the same way that brain damage involving the motor cortex will cause a paralyzed limb, damage to the areas of executive function will also cause disability, but of a different sort. In this case it changed the fundamentals of Mike's personality and judgment. Mike's behavior in my clinic was what neurologists refer to as "frontal." He was inappropriately exuberant for the setting. He lacked empathy and could not interpret or respond to his parents' distress. He perseverated – said the same thing over and over. He was inflexible. When I tried to change the conversation he could not adjust.

If Mike was sexually inappropriate to a stranger in the park it was not his fault. His behavior was a symptom of brain injury. That is not to say that any trauma suffered by the victim of his attention should be dismissed. But maybe it would help the stranger to know why Mike had behaved as he did – or maybe it would make no difference to them at all. I wrote a medical report and sent it to him to share with his solicitor. Meanwhile I still had the problem of Mike's unexplained bruises and bitten tongue to solve.

Mike's brain contusions were an excellent substrate for seizures, but nobody had seen Mike have a seizure. Mike's own account of what happened at home couldn't be trusted. A court case loomed but in the meantime Mike had been released by police and was living under the supervision of his parents. I hesitated before arranging to admit Mike to the video telemetry unit for observation. I wasn't sure that he would tolerate the confinement and was concerned about how the ward would manage. I discussed his behavioral issues with the senior nurse and we decided to go ahead with the admission. Anticipating problems didn't stop them, however. Mike got in a heated argument with another patient within hours of arriving on the ward. I was asked to discharge him for the safety of other patients and the staff. And for Mike's own safety and well-being.

I asked the technicians to stick the electrodes to his head before he left. He could be sent home wearing them and carrying the headbox in a bag worn across his body. A member of his family would bring him back every morning for the recorded information to be downloaded and for the electrodes to be reattached. If he collapsed in that time we would at least have his brainwaves to study.

"If we see anything we'll video it," his mother told me.

"That would be immensely helpful. Thank you."

Mike didn't collapse that week. The test was normal. It was hours of work and we were no further forward. All the time I kept thinking that whatever was causing his injuries seemed the very least important thing going on for Mike. I was stuck on a treadmill that doctors often find themselves on. I knew I couldn't change Mike back to how he had been. In desperation I was still looking for something I could fix. As if finding and eliminating seizures would get Mike his job back, or his girl-friend or his old lifestyle or eradicate the court case. It would not do any of those things.

*

Most people I have met who have suffered a severe head injury have a similar thread to their story. A severe head injury is one in which there has been a prolonged loss of consciousness, a fractured skull, a brain hemorrhage or an injury that is accompanied by significant amnesia.

The changes that brain injury bring are often very difficult for the person to appreciate themselves. One young man I know who worked as a teacher was certain that he could still do his job once the worst of his traffic accident appeared to be behind him. He, like Mike, felt fully recovered once the visible wounds had healed. He had been in the same school for five years, so he knew his job well. Lesson plans were clear in his head, but in front of a class he lost control. He got frustrated and upset and angry. It took a year of disastrous lessons for him to understand that he could no longer process several sounds at once. His temporal lobe was damaged, and auditory processing takes place in the lateral temporal lobe. Any more than one voice in the room was too much for him. Such a disability is not one we can test for in a clinic and it's hard to describe or understand if you are the person affected. The teacher believed that he was fully recovered and so did everybody else. And much of the time he was perfectly well – until he was faced with a class of excitable noisy sixteen-year-olds where he was instantly invalided.

I once heard a victim of head injury speak at a conference. He had come to talk to a room full of neurologists and neuro-surgeons about his personal experience of head injury. He had been a London banker. A successful man pre–head injury, he was now building a new life from the new version of himself. He had suffered a traumatic brain hemorrhage in a skiing accident. The venue where he was speaking was a semicircular lecture theater. He stood behind a podium at the center of the semicircle. What he did not realize and could not have realized is that he had directed the entire lecture exclusively to the right side of the room. He gave them all his attention, while those sitting to his left might as well not have existed. This is a feature of something called "neglect." Strokes and head injuries that cause damage to the right parietal lobe can cause a person to become largely oblivious to their left side. Not just of their environment, but also their own body. Their left arm doesn't exist. They could put on half their jacket and ignore the rest. Shave half their beard. They might eat only the food on the right side of their plate. To be reminded to eat the rest somebody would have to turn the plate around so that the second half would enter their field of perception. There is nothing wrong with their vision, but rather with their attention. Most importantly, much like Mike's experience, they have no insight into their own disability. They cannot appreciate their own attentional deficit until it is explicitly pointed out to them.

Mike struggled with insight. I imagine it contributed to the fact that he stopped turning up to clinic and didn't reply to letters. He couldn't appreciate that he needed medical help more than ever. I suppose he didn't regard his arrest as a medical problem. I didn't see Mike again for a long time after his arrest. The next I heard from him was when his father rang me and asked for advice. Mike was intransigent. His own worst enemy. I rang Mike. He was a reluctant participant in the conversation. I persuaded him to attend the head-injury service. I appealed to the lawyer in him. It seemed to work. He could not understand why his behavior was upsetting people, but he knew what a court case was and he knew the seriousness of his position. He had seen his life slip between his fingers. I reiterated that the head-injury service would try to get some normality back for him. I knew I was exaggerating. A life without a criminal conviction was what I was really aiming for.

It took several months of assessment and advice from the specialists in the aftermath of his head injury for Mike to appreciate his new limitations. Work with a psychologist drove home the fact that a full recovery was impossible. It also allowed him to play with ideas about how to build a new life. With encouragement he came back to my clinic after a year's hiatus. When he came through the door I could see an improvement immediately. He was lighter. His parents were with him and he accepted their presence this time.

"Dare I ask, is the issue of the court case still hanging over you?" I said once we had been through the preliminaries.

"She dropped the charges!" Mike said clasping his hands together like a prizefighter.

"She withdrew the complaint," Mike's father clarified. "The lawyers met and I went along. We brought all the medical reports and discussed it rationally. The woman said she was satisfied and that she had no desire to take it further."

"I sent her a letter of apology," Mike said.

"She didn't ask for one," Mike's father said, "we passed it on to her lawyer."

In an odd way the arrest proved to be Mike's savior. It forced him to reflect on how things had been since the accident. He could see that people were responding differently to him. The psychologist helped him link consequences with actions. She also supported his family. They had had to stop expecting Mike to go back to how he had been. Everybody needed to change.

"We've had to just accept that he's a different Mike than he was," his mum said. "It's been a matter of getting used to that for us too. Actually there are ways in which he's a nicer Mike.

He's funnier. He's less driven and more affectionate. We see a lot more of him. That's not to say that I love the fact that he blurts out everything that comes into his head."

"Yes," his father added, "if you want a healthy dose of the truth go to Mike!"

There had also been a breakthrough in finding the cause of Mike's mystery injuries. This was also facilitated by Mike's arrest. He had been forced to move home with his parents, which meant that his mother was present to see him have a seizure. One evening he was playing video games when he collapsed. His mother heard him fall and ran to find him lying on the floor convulsing.

It was easier for me to treat Mike's seizures with confidence once the diagnosis had been confirmed. Until then I didn't know if I was throwing epilepsy drugs at fresh air or not. I asked Mike to take a higher dose than I had previously prescribed. I was relieved when it worked. Although I was quickly reminded that you should never rejoice too soon in the treatment of unpredictable capricious illnesses. Mike went eight months with no seizures and then had three in a single month.

"He was really getting back on track," his mother told me.

Mike had started doing supervised work at his father's firm. He could manage one clearly defined project at a time. Every job he was given needed to be presented as a list that he could check off once each part of the process had been completed. Both his home and his workplace became ordered around routine and quiet. The return of the seizures had disrupted that progress. It was only when I was just about to change his epilepsy drugs that the reason for their return made itself known.

A blood test taken to check the level of epilepsy drugs in Mike's blood came back negative. If he was taking his tablets there was no evidence of it.

"I didn't think I needed the tablets anymore," Mike told me when we discussed it.

He had abruptly stopped taking his medication of his own volition.

"You'll always need them, Mike," I told him. "The scarring in your brain will not be going away and therefore neither will your epilepsy. But your seizures might go away – if you keep taking the tablets."

Taking his medication was another thing that needed to be formally scheduled and recorded in Mike's life. "You were not clear enough about the tablets. You never said I had to take them forever," Mike corrected me in no uncertain terms. "You will really have to do better, Dr. O'Sullivan!" he said, and laughed.

Quite right.

11

ELEANOR

Medicine is a science of uncertainty and an art of probability.
—William Osler (1849–1919), physician

At 6.30 p.m. on Tuesday, December 16, 1997, 4 million Japanese children sat down in front of the television to watch a cartoon. The program was called *Electric Soldier Porygon* and was one of a series of Pokemon cartoons screened at that time every week. About twenty minutes into this particular show the well-known character Pikachu blew up some missiles with a lightning attack. The animators used blue and red strobe lighting to create the illusion of an explosion. It was at this exact point that children viewing the show were said to have been taken ill. Some only felt dizzy or sick, but a few lost consciousness or had a full-blown convulsion. The emergency services found themselves inundated with an unprecedented number of phone calls. Ambulance crews took 685 of the affected children to the emergency department. One hundred and fifty of those needed to be admitted to hospital. Later that evening a news program reported what had happened. During the report they showed the cartoon sequence. The phone calls to the emergency services started all over again ...

We have all heard the announcements that are made during news broadcasts, "This segment contains flash photography." That epileptic seizures are triggered by flashing lights is one of the things that people think they know about the condition. It is true of course. A certain proportion of people with epilepsy will have a seizure in response to flashing lights. It is one of the reasons that people newly diagnosed with the condition are frightened to use their computer or sit too close to the television. Some even worry (unnecessarily) about the flickering fluorescent strip lighting of supermarkets. But the truth is that most do not need to worry at all. Less than five percent of people with epilepsy are photosensitive, and of those, they are usually children. It's unlikely that every one of those children who felt unwell during that Pokemon cartoon had epilepsy – reports raised the possibility of contagious hysteria as a contributing factor – but in a small number of them it was established that they had had a photosensitive seizure.

Photosensitivity is a real but exaggerated feature of this disease. But as with so many facts about the brain, the truth is stranger than the rumor. Epilepsy, in particular, is all too often odder than the folklore that surrounds it.

*

Eleanor was nine years old when her mother first noticed that she was becoming a clumsy child. Eleanor loved school and was bright and talented. She also loved dancing. She had been a good dancer, but suddenly she was uncoordinated and began to stumble. In the midst of a plié, something she would have mastered elegantly before, she fell over. Eleanor came from a large family. She had two older sisters, as well as an older and a younger brother. Her parents knew that this abrupt change in their youngest daughter was something more than just the result of an ungainly pre-adolescent growth spurt. They took her to the family doctor but he found nothing wrong. Eleanor's mother even asked her daughter to demonstrate the basic ballet positions to the doctor in the hope that it would reveal the problem. Eleanor refused.

"I remember being horrified. So embarrassed," Eleanor told me when she recounted the incident many years later.

The cause of Eleanor's problem proved very hard to pin down. Most of the time she was quite well. Only when she was engaged in activity – sport, ballet – was it evident that there was something wrong. She didn't collapse or lose consciousness, but she lost her balance and tended to veer from side to side.

Eleanor was too young to really appreciate that she might have a medical problem. However, having seen their older children grow up without any difficulties her parents were convinced that she did. Every time they took her to the doctor she defied them by appearing the epitome of health.

Eleanor was a little more worried when the loss of balance turned into falls. Especially when she fell over several times in a single week and was forced to stay off school.

"It got so bad at one point that I couldn't even walk. I didn't know what was stopping me and I do remember being upset by that," Eleanor told me.

This significant deterioration prompted a referral to a neurologist. It took a week for her to get an appointment. By that time she was better. The neurologists examined her. They arranged for her to have some basic tests. The tests were normal and she was discharged from clinic.

For the next eight years Eleanor was in a cycle. Every six months or so she had spells during which she blundered into walls and fell over randomly. Doctors kept telling her parents that she was becoming a graceless and gawky teen and that she would grow out of it again. Her parents were convinced the doctors were wrong. They knew her. They saw the change in her from week to week. Feeling hopeless, Eleanor's mother took her to a healer.

"He put his hands on my head," Eleanor told me. "He laid hands on me!"

"Did it help?" I asked.

"I'm here aren't I! There was no miracle," she answered.

She laughed when she recounted this, but it made me wonder how desperate her parents must have felt to have taken that route. Both her mother and father were scientists and I guessed they would not normally have been inclined to extreme alternative therapies or superstition.

"How did they finally figure it out?" I asked her.

"Professor F made the diagnosis."

By the age of seventeen Eleanor spent one week every year virtually bed-bound. Her GP referred her for multiple medical opinions until he got an answer. A neurologist asked the family to video what she looked like when she couldn't walk.

"Mum brought a video to Professor F," Eleanor told me. "I remember being furious. I didn't want her to film me, but she did. He looked at the video and said it was epilepsy."

Learning to recognize how diseases behave can trump test results. Eleanor's clinical examination was normal. Her tests were normal. Eleanor's family videoed her as she spontaneously and unceremoniously fell to the floor. An experienced professor of neurology immediately suspected epilepsy. He put her on treatment. I met Eleanor another eight years after that.

"What's been happening over the last few years?" I asked her, knowing it had been a long time since she had gone back to see the doctor who originally diagnosed her epilepsy. "Why haven't you been going to the epilepsy clinic?"

"I didn't think I needed to," Eleanor told me.

"Until recently she'd been doing okay," her parents confirmed.

"Are you still taking epilepsy medication?"

"I'm taking lamotrigine, but I'm not sure it's helping."

After Eleanor was diagnosed she had been put on a low dose of an epilepsy drug. Taking it every day, twice a day, had felt unnecessary to her. She didn't fall every day. The attacks came in short clusters every six months or so. It was like having a bad cold twice a year – an irritation that she dealt with when it arose but didn't worry her at other times. The epilepsy drug didn't change that pattern so Eleanor didn't pursue further treatment. She gradually stopped going to see the neurologist.

"So why are you here now?" I asked.

"A weird thing happened to me. I got a job on a cruise ship and struggled to walk while I was on the boat. I had to leave the job in the end. I want to know why."

Eleanor had found dancing too difficult to pursue as a career because her balance was so unpredictable. Sometimes she could dance and sometimes she couldn't. She continued to love to perform and to be in the limelight and she did an A level in drama. Then she went to college and took a performing arts course.

"I want to go into production," she told me, "I took the boat job just so that I could travel."

Eleanor had taken the post of entertainment hostess on a Caribbean cruise ship. The timing of her start date was not good. The weather was unseasonably bad and the seas were rough.

"Everybody felt sick, all the staff. Not just me," she said.

It was true. All the new staff on the boat struggled with seasickness for the first week of the trip. However, Eleanor's problem was different to everybody else's. Even when she wasn't feeling nauseous she couldn't walk in a straight line. Along the narrow ship's corridors she bounced from wall to wall. In the end she had had to hand in her notice, disembark and fly home.

"How far did you get?" I asked her.

"I got off at Saint Lucia."

"Not too bad a place to end up."

"I was really disappointed, but I just couldn't stay on the boat a moment longer."

"It sounds as if you're describing a balance problem rather than seizures," I told her.

I asked Eleanor to stand up so I could watch her walk. She walked normally. I asked her to walk heel to toe as if she was on a tightrope. Then to walk on her tiptoes and on her heels. I checked her reflexes and her coordination. Everything was as it should be.

Her examination was normal. That didn't matter. It was the story. A more senior neurologist had heard epilepsy in it, but I just couldn't. Epileptic seizures usually have a tempo that I recognize. Seizures are typically brief attacks that last seconds or minutes. A transient lightning strike. Eleanor seemed to have days of being unable to walk. The period of feeling unwell seemed too long to be explained by a seizure. No less than any other type of detective, a neurologist must question the story they have been told if something about it doesn't sound right. My mind ticked over with associations. I didn't know what was wrong with Eleanor, but she made me think of somebody I had met before.

*

I had looked after Emily when I was still a trainee neurologist. Like Eleanor, she was a young woman who took to her bed every now and then. She had been referred to the consultant I worked for with a letter containing a suspicion that there wasn't much wrong with her. As if being bed-bound could ever be dismissed as nothing. I had met Emily on the neurology ward after the consultant had asked that she be admitted for tests.

"About once a month my whole body turns to lead," Emily told me. "I wake up in the morning and I feel like I have a truck sitting on my chest. My arms and legs just aren't strong enough to lift my body."

Emily's story was very memorable, not least for how it started. For her tenth birthday Emily had a party in her garden. Emily was allowed to have a party only every third year, alternating with her brother and sister. It was exciting for her. The party itself was a success, but the next morning Emily's mother had great difficulty getting her out of bed. When Emily did get up her walking was strange. Her mother allowed her to go back to bed and let her sleep a little longer. She was worried and would have called the doctor except that when she went to wake Emily a second time she seemed to have made a full recovery.

After that, according to Emily's family, she became allergic to birthday parties. And tiredness. And exercise. And junk food. Exposed to any of these things, Emily was likely to take to her bed. At the start it had lasted only for an hour or two, but in time some attacks lasted days. She, like Eleanor, saw lots of doctors. She found herself labeled as attention-seeking, something which angered her and her family. One doctor thought she might be allergic to food additives and another that she had a gluten intolerance. Most didn't really know what was wrong with her.

The easiest way to make a neurological diagnosis is to follow the symptoms to their anatomical source. Rather than worrying about *what* the pathology may be, it is more sensible to follow the clues to *where* it is in the nervous system and then focus the investigations there. That distinction can usually be made on a clinical examination. If somebody has a weak arm then you search the anatomical pathways that allow that arm to move, from the fingertips to the brain.

But some neurological problems are so rare and transient that this isn't possible. Then you have to spot the patterns and try to match it to something you have heard before. The description of Emily's condition had immediately struck a chord with the consultant I worked for. He brought her into hospital to test his suspicions.

There are a small number of exceedingly rare conditions that cause intermittent paralysis in young people. Electrical activity is as important to the way muscles work as it is to the brain. Contraction and relaxation of muscles depend on the difference between intracellular and extracellular concentrations of sodium, potassium, calcium and chloride ions. At rest, an imbalance of these ions is maintained by gates in the cell wall, referred to as channels. Channels open and close in response to chemical messages from the nerve ending. Ions are electrically charged and their movement in and out of cells changes the electrical difference across the cell membrane. This movement of ions is essential for normal function of the muscle. If any of the ion channels do not work properly then the muscle may not be able to contract, and can become weak and flaccid. In one such condition (called a channelopathy), carbohydrate-rich foods result in potassium being driven into muscle cells. If a person eats large amounts of such foods they can develop transient paralysis.

That is what Emily had, a genetic condition called hypokalemic periodic paralysis. Birthday parties made her sick because party food tends to be carbohydrate-rich. It is a problem so rare that only one in 100,000 people have it.

*

How many people does a doctor meet in their career? I don't know, although the odds were that, having met Emily, it was unlikely that I would ever meet somebody like her again. But now I thought I saw her in Eleanor. Something on that boat was making her sick. What was it?

"Was your diet very different when you were on the boat?" I asked Eleanor.

"Not really," she said, "I know I'm silly but I worry a lot about what I eat. I'm quite careful with my diet."

"She is a very careful eater," her mother confirmed.

"Were you working very irregular hours?"

Perhaps she had been sleep-deprived. Or was she stressed?

"I worked late nights but I didn't have to get up early so I wasn't overtired. I don't think I was anyway. And I definitely wasn't stressed. I loved the work."

I couldn't figure it out and I told Eleanor so. But I would try. I arranged the usual scans and EEG tests and they were as normal as I expected them to be. Tests done to look for muscle problems similar to Emily's were also normal. That was okay – these investigations were only a baseline to be compared with similar investigations whenever her symptoms came back. If Eleanor's attacks continued to behave as they had for the previous eight years then they would happen again. All I had to do was wait. Eleanor's walking problem lasted for ten days at a time. The next time it happened she would ring me and I would admit her to hospital, video her and repeat all her tests.

A few months later I got the call. I immediately admitted Eleanor to the neurology ward. She arrived with her mother and father supporting her. I could see straightaway that she wasn't the Eleanor I had met before. She was certainly unsteady on her feet. Her parents held on to her as she walked. I asked them to let go so I could see her movement better. Eleanor refused to allow it.

"I'm scared," she said.

"I'll be beside you, I'll catch you if I have to," I reassured her.

"No," she said firmly, "you won't be quick enough."

Watching Eleanor I thought how like a marionette she looked with her parents holding an arm each and helping her to her bed. As her mother took off her coat she let go for a moment but she kept herself ready to step back in to grab Eleanor if needed. I couldn't decide if the vigilance was necessary or not.

Eleanor wouldn't stand up so I examined her lying on the bed. In that position her arms and legs were very strong. I asked her to run her heel quickly up and down her shin, right and left in turn. I asked her to touch her nose and then to move her finger between her nose and my upheld finger. I asked her to pretend she was playing the piano. Although standing up she had been fragile and nervous, lying down she was confident and her movements and coordination worked seamlessly. I could find nothing to explain her fear of standing. Was the fear greater than the reality?

"Move around the room as much as you can," I told her. "It won't help if all I can see is you lying in bed all day."

"I'll do my best, but I'm scared to walk. Can I call the nurses to help me?"

"Of course, if you need to. How do you get to the toilet at home?"

"Mum takes me. And I have a bedpan I can use at night if I need it," Eleanor looked very embarrassed to have to admit this.

"Are you sure I can't ask you to walk on your own so that I can see what happens?" I asked her. I had no sense of what all these precautions were protecting her against.

"If I have to ..." Eleanor said and seemed to begin to steel herself to move from where she was reclining against some pillows. After some uncertainty she sat forward. She had no sooner done so than she fell backwards again onto her pillow.

"What was that?" I asked. I couldn't tell if she had lain down deliberately or had fallen.

"That was it. That was a seizure," Eleanor told me.

It had been too quick for me to get a sense of it. She hadn't lost consciousness and there was no involuntary movement. She just sat up and dropped right back like a rag doll.

"That's a start," I said, confused. "At least I know what I'm looking out for. But press the alarm every time it happens, won't you? It's so quick, I don't want to miss it."

"I will. If I can."

"I'll be here. I'll do it," her mother said. "Can I stay with her overnight too? I'm really worried about how she'll get to the bathroom."

Being trapped in the video telemetry room is worse than most hospital stays. It is very restrictive. In acknowledgment of that the ward staff let Eleanor's mother stay outside of normal visiting hours. The technicians came to wire Eleanor up and we waited. The next morning I came in and set about looking at the previous day's recording. As usual I started by scrolling through the events list on the computer to see if Eleanor had pressed the alarm button at any stage. She had pressed it over fifty times. I clicked on the marker for the first button press and then scrolled the video back a couple of minutes.

Eleanor was lying in the same position I had left her in. She was supported fully by her pillows and was looking slightly to the left where her mother sat in a chair. They were chatting and laughing. She looked relaxed and happy. On the video I heard a knock on the door and Eleanor lifted her head ever so slightly to look towards the door, which was to her right. She was only halfway through doing this when her head, which had risen only an inch off the pillow, fell back again. Her mother reached over and pressed the alarm button.

"Damn," Eleanor said and beat the bed with her hands. After a moment she lifted her head again, but this time without difficulty. She greeted the nurse who had come through the door to check on her. Another nurse who had heard the alarm followed the first into the room.

"Did you press the alarm or was it a mistake?" the second nurse asked. She walked closer to the bed and examined her. "You look okay. Did you have a seizure?"

"Yes, but it's gone. They're very quick."

I clicked on the next event marker on the computer. This time Eleanor had extra pillows behind her back and lay in a more upright position. A tray-table had been pushed in front of her and she was preparing to eat her lunch. I watched as she put her hands by her sides to push herself a little further forward to make eating easier. As soon as she tried to do so her whole body flopped back against the pillow. It was very brief. A flash. There was no obvious loss of consciousness and again she groaned and seemed frustrated. As soon as it was over she put her hands down by her side again and this time successfully adjusted her position and began to eat her lunch.

I moved from one button press to the next on the computer menu. Most were the same as the first two. One that differed slightly happened as Eleanor tried to lift a forkful of food. Halfway through doing so, her arm dropped limply by her side. The fork slipped from her hand and landed beside her on the bed.

I started to look for an attack that happened when Eleanor was out of bed. It was hard to find one. She had stayed lying in almost exactly the same position all day. I was several hours into the study when I finally hit on a near fall that had occurred while Eleanor was trying to stand. It seemed she needed to go to the bathroom and had no choice but to move. Her mother was with her but they both seemed nervous so they had called a nurse to help. Very slowly and gingerly Eleanor's mother had maneuvered her until she was sitting on the side of the bed. The nurse was on one side and her mother on the other. There seemed to be a great deal of building up courage before Eleanor attempted to come to a standing position. I was struck again by the apparent excess of caution. But I quickly learned why it was necessary. Almost as soon as Eleanor stood her whole body went limp. She couldn't support herself. She began to fall forward but her mother pushed her briskly and changed her direction so she fell backwards onto the bed. She ended prostrate on her back.

"Bloody hell," I heard her exclaim.

She was clearly wide awake and irritated to have her trip to the bathroom interrupted in this rude way. The fall onto the bed lasted less than a second. Afterwards she immediately righted herself. Then she restarted her journey, but only after ensuring that both her mother and the nurse were holding her tightly. She took the few steps across the room and then disappeared from camera view. I heard her let out another small cry from the bathroom. I could only assume that was an indication it had happened again.

By the fiftieth attack I could see that the elements of each were the same. Eleanor tried to move and as soon as she did so she lost muscle tone in her whole body leading to collapse. Her consciousness was not affected.

I looked at the EEG. When Eleanor was well her brainwaves looked quite normal. In this record I could see a distinct change in the background pattern immediately before every collapse. The change was not the sawtooth pattern that occurs in many focal seizures, but rather a spindle of tiny waves situated just over the vertex (the apex) of the skull. It was not on one side or the other, but right in the middle of the head. Eleanor's attacks were not conventional seizures but the EEG discharge confirmed that epilepsy was the only explanation. I should never have doubted it.

Seizures are a unique disorder in the way they can produce either loss of function or excess function. One person loses their speech, and the next has forced involuntary speech. One person loses their vision, and the next sees a complex hallucinatory scene. Muscles may become stiff and jerk if activated, or alternatively they may become paralyzed and lose both strength and tone. Eleanor lost function. She developed atonia – complete loss of tone that caused the muscles supporting her posture to relax to such a degree that she couldn't support herself.

Muscle tone is the continuous partial contraction of muscles required to keep us upright. The frontal lobes control many aspects of movement. The primary motor area exerts the simplest level of control over voluntary movement. The supplementary motor area and premotor areas of the frontal lobe are more concerned with planning and spatial awareness. They are also involved in stabilizing posture. Seizures arising in the supplementary motor area in particular have been associated with loss of postural tone. This seemed to be Eleanor's problem. Every muscle in Eleanor's body lost tone simultaneously. It was brief but long enough to cause her to fall. With only the muscle tone affected the seizure did not cause her to black out.

As I watched the video other things about the seizures struck me. She did not have any attacks when she was relaxed on her bed, either chatting to her family or watching television. She only had them when it mattered – when she decided to move. Somebody knocked on a door and she turned her head – and had a seizure. She lifted a fork and had a seizure. She tried to stand and had a seizure. Her seizures were triggered by movement. Reflex seizures.

My thoughts wandered back to the cruise ship. Eleanor could walk on dry land. The Saint Lucia beach presented no challenge to her. The long corridors of a swaying boat on a choppy sea caused her to fall constantly. It was the movement of the boat and what that required of her postural muscles that was the problem.

Reflex epilepsy is a group of particular epilepsy syndromes in which a specific stimulus triggers a seizure. Photosensitive epilepsy is the commonest reflex epilepsy. The rarer sorts only lend further mystery to the brain.

Hussain is one of several patients I know who has seizures when he eats. It doesn't happen every time he eats, but a seizure almost never occurs if he is not eating. It is common enough that I have seen several cases of this "eating epilepsy." In the video telemetry unit if a person says they only have seizures in response to eating a pepperoni pizza then we will provide a pepperoni pizza to try to trigger an attack. Seizure triggers are strange and I will believe pretty much any pattern that a person tells me. We experiment with almost any trigger that is suggested to us – alcohol, the smell of perfume, fights with a mother-in-law, sunlight through a curtain, a sudden noise.

Philip's account of his seizures was more unusual still. He has seizures in response to hearing a particular sort of music. Once he was in a coffee shop when acoustic guitar music abruptly blared through the speakers. Immediately Philip collapsed and had a convulsion. This is called "musicogenic epilepsy." In it the brain cannot seem to tolerate particular types of music, or frequencies, or a specific pitch. Sometimes just thinking about music is enough to provoke a seizure. That overlaps with an even rarer phenomenon referred to as "thinking epilepsy." Trying to solve a Rubik's cube, playing a particular board game, reasoning through a problem, even thinking a certain thought: each of these has been reported to trigger a seizure.

Patients with these reflex seizures possibly have regions of cortical hyperexcitability that overlap or coincide with areas physiologically activated during specific sensory stimulations and cognitive or motor activities. When these areas receive the appropriate stimuli (reading, thinking, moving, eating) a critical mass of cortex is activated and epileptic activity is provoked.

Was Eleanor's odd story an example of a reflex epilepsy? An atonia triggered by movement? The answer existed in watching Eleanor's week unfold. I kept recording to see what other information became available. Every day she had more seizures than the day before. By the end of the week it was clear why Eleanor spent one week a year never leaving her bed. By day five she was having 500 seizures a day. She couldn't move at all. The short walk to the bathroom became an impossibility. In the hospital the nurses pushed a commode right up against Eleanor's bed. When she needed to use it she alerted the nurses. They tactically redirected the camera towards the ceiling and helped to support her as she made the transfer. I was wondering how she managed even that small journey when I noticed another pattern. When Eleanor had to change her position to sit on the side of the bed she became anxious. She began preparing herself for movement. She moved and simultaneously had a seizure. But then immediately after the seizure there seemed to be a minute or two in which no seizure would occur and that gave her a vital window of opportunity to get things done. Soon I realized that Eleanor sometimes gave herself a seizure deliberately. She moved, had a seizure and then in the post-seizure period she did whatever she had really needed to do. Lift her fork to her mouth. Transfer to the commode. Make herself comfortable in bed.

I had doubted Eleanor's need for such extreme caution when she walked onto the ward with her parents. I soon felt guilty about that. Her fear was more than justified. As her seizures crept up to hundreds per day it was obvious that if she was not very careful she would come to serious harm. She never lost consciousness and the attacks were very short, but such a complete loss of motor control rendered her very vulnerable. On one occasion I saw her lift a sandwich to her mouth and take a bite. Just as she did her whole body went limp. She fell backwards and was caught by her pillows. Her hands dropped by her sides but somehow the sandwich stayed flopping around in her mouth. Her mother, who spent most of the day by her daughter's bedside, had looked away for a moment. She turned back just in time to grab the bread and hold it in position until Eleanor recovered sufficiently to finish biting and swallowing. On another occasion Eleanor had a bowl of soup on the tray-table in front of her when she suddenly fell forward. Her face hit the table, landing inches from the bowl. She suffered a scratch to her forehead but was mostly just relieved to have escaped a scalding. After those incidents mealtimes were much more closely supervised.

By the end of the week Eleanor did not even consider trying to stand.

"I think we urgently need to do something about this," I told her.

Status epilepticus involving a generalized tonic clonic convulsion puts the person affected in immediate danger of death. If emergency drugs don't stop the seizure then the person must be sent to the intensive care unit. There, their muscles are paralyzed, their brain is anaesthetized to suppress the seizure activity and their breathing is taken over by a ventilator. But status epilepticus can also be a cluster of smaller seizures between which the patient doesn't recover fully. Eleanor never lost consciousness so her seizures were not as life-threatening as a prolonged convulsion. Still there were so many per day that they represented status epilepticus and warranted considerable concern.

"I want to escalate your epilepsy drugs to see if we can get this under control," I told her.

In the past Eleanor had not sought treatment when she found herself almost entirely immobilized. She waited out the problem. I couldn't countenance only watching as she had so many seizures per day. I hoped to bring the cluster to an end with drugs. Eleanor was uncertain.

"I don't know if I want to take any more medication," she told me.

Eleanor had remained on a very low dose of a single drug for many years. It had never helped.

"You might escape injury hundreds of times with these seizures but you are unlikely to be so lucky every single time. And I'm worried about the toll that so many seizures are having on your brain."

Usually a person in status is unconscious. Their life is in danger so emergency treatment is mandatory. Eleanor was wide awake and having a rational conversation with me about how to proceed. But she was confined to her bed, reliant on bedpans and constant supervision. It was a surreal situation for us both. I asked her again to consider taking emergency medication. She agreed.

First I gave her tablets in escalating doses. The seizures continued. When things got worse instead of better I set up an infusion to be given intravenously over thirty minutes. This sort of treatment is supposed to act within minutes. It had no effect at all. Every day a technician or I counted the seizures. They stayed at a rate of hundreds per day. They were all the same – a complete flop like a rag doll in response to any attempt to move.

Eleanor's admission was not a success. The diagnosis of epilepsy had been confirmed beyond doubt and it had been refined. I could surmise that the seizures were arising in the supplementary motor area – but knowing that wasn't helping me make her better. In fact this cluster of seizures was longer than she had ever had before. Like many people with epilepsy I had to treat Eleanor without fully understanding her problem. The discharge on her EEG was the only objective evidence that something was wrong with her brain. It was proof of epilepsy, but the discharge was ill-localized and the brain scan was normal. I had only what I had observed and comparison with case reports in the literature to go on.

A sixteen-year-old boy in Pennsylvania had tonic stiffening of one arm and leg that were triggered by rope-climbing, push-ups, running and most sorts of exercise. All his tests were normal. Epilepsy was the presumed diagnosis but he had failed to get better with treatment. A sixteen-year-old Italian boy developed stiffening of one arm in response to movements of his fingers. Searching for coins in his pocket was sure to set it off. Scans suggested an area of possible abnormality in his right frontal lobe. Epilepsy drugs successfully brought his seizures under control. A thirty-five-year-old man from London developed stiffness and jerking of one limb in response to stretching exercises – made worse by mental stress and emotional embarrassment. Medication did not help him. An exploratory operation revealed a scar in the supplementary motor area of the frontal lobe on the left side of his brain. Removing the scar left him with transient weakness in the right arm. But it also got rid of the seizures.

Nobody else that I know of has Eleanor's exact sort of epilepsy. Even if they did, no two people would have the same personal experience of it, and people's reactions to drugs are very different. Eleanor and I were both learning as we went along. I was trying to find a treatment that suited her and she was showing me how a person overcomes a disability so unique that the people caring for her don't know what to advise. Eleanor learned the dangers and how to avoid them. She accepted the inevitable seizures and used the calm between brainstorms to make as much progress as possible. She learned what was safe to eat and which sort of hot food she could not have in her vicinity unless she also had strict supervision.

After a while the lack of progress became obvious to us all.

"I think the drugs are making me even worse," Eleanor said to me one day.

"Epilepsy drugs can make seizures worse," I confirmed, "but I don't think we can stop them and just do nothing. I think we have to keep trying new things and hope that the next one works better."

Eleanor, who had been on board with the treatment, albeit after persuasion, was becoming more reluctant. Her parents were more reluctant still.

One day I watched as a nurse asked Eleanor to try to sit in a chair for a while. Stuck in bed Eleanor was at risk of pressure sores and blood clots. She needed to move around or at least change position. I saw that Eleanor was very scared to move and needed a lot of support. As I watched I saw the seizure start just in the planning phase of movement, before the movement itself had even begun. The supplementary motor area is implicated in just this planning of movement. From what I had seen I became concerned that Eleanor's seizures were worse because of the degree of fear and anxiety that now surrounded every movement. I thought of the case report of the man whose seizures were made worse by mental distress. I wondered if I was making her worse, not through giving her drugs, but through the stress of this admission.

I decided to take a different tack. I agreed to stop the epilepsy drugs. She was happy; it was what she wanted. Whether or not it was the epilepsy medication that was making her worse was a matter for debate, but it certainly was not making her better. I began to focus on the fear and anticipation of movement that seemed as likely to bring on an attack as the movement itself. I asked the physiotherapists to try to take her for walks. I hoped to build her confidence again. I also asked the psychiatrist to address the fear of walking. While it was a justifiable fear it could also be slowing her recovery.

Eleanor had been admitted for a seven-day period of video telemetry monitoring. In the end she stayed for seven weeks. Nothing I did, neither drug nor psychological intervention, made any positive difference. It was certainly the longest cluster of seizures she had ever had. In the end it was only time that seemed to make her better. The cycle just wore itself out. The seizures got fewer and fewer. We both agreed it was in spite of, rather than because of, anything I had done. I know Eleanor and her family think I made her worse by giving her epilepsy drugs. I agree I made her worse but I think I did it by creating mental anxiety and giving so much attention to the attacks. If movement and eating and flashing lights and music and thinking a certain thought can trigger seizures then surely anticipation and anxiety could do the same.

That all happened more than ten years ago. Despite my failure, Eleanor has stayed under my care. More than once I was so worried that everything I was doing for her was wrong that I referred her to other specialists or discussed her case with them. So far nobody else has come up with a solution. The treatment I have offered, however inadequate, is all there is.

Unfortunately Eleanor's condition has progressed. Her seizures no longer restrict themselves to biannual clusters. They have started to happen spontaneously, rather than just with movement. She also has convulsions now. This means that the electrical discharge is no longer restricting itself to a small patch of frontal lobe but has found a path that allows it to spread to the whole brain.

I keep reinvestigating and re-evaluating. In about 2014 something happened to give a teasing glimpse of a hope for change. A new-generation MRI scanner finally revealed an abnormality in Eleanor's brain. Sitting neatly in the left supplementary motor area is an unwanted benign tumor of the sort that is present from birth. We had always suspected it was there but now it was visible and it fitted perfectly with the theory of where her seizures were arising. An explanation for seizures does not equate to a cure but it opens avenues. It was a relief to have proof.

That discovery set Eleanor and I off on a long tedious pursuit of the possibility that Eleanor's seizures could be cured by surgical removal of the tumor. The MRI and clinical theory fitted together. For a person to be suitable for surgery the two most important factors are the need to identify exactly where the seizure is arising and to ascertain that the surgical target area is not in the eloquent brain.

Eleanor had an fMRI scan that looked at her brain activity when she was asked to tap her right hand and then her right foot. Blood flooded to the hand and foot regions of the brain and that was reflected on the scan. The activated area of brain that represented the motor region for the foot sat treacherously close to the tumor. Almost on top of it. When the problem was discussed with the surgeon she thought that the operation might cure Eleanor's epilepsy but leave one leg paralyzed. She might go from being in a wheelchair for one reason to being in a wheelchair for another.

Endlessly brave and forward-looking, Eleanor agreed to have an intracranial EEG study to delineate the risks more completely. The surgeon opened Eleanor's skull. She placed an array of sterile electrodes directly on the surface of Eleanor's brain, over the tumor and in the area around it. Then Eleanor's head was covered in a bulbous sterile bandage. With wires trailing through bandages that had exposed brain lying underneath, Eleanor returned to the video telemetry unit for monitoring.

"I hate this," she said when I called to see her, "I'm not used to being unwell."

I felt my heart skip a little. Eleanor's comment seemed to consolidate something I had always thought about her. Eleanor had seizures every day. She had to change her life drastically as a result. But she did not consider herself ill. She accommodated her epilepsy but it didn't define her.

After having left her job on the cruise ship and returned to the UK, she had worked in a local restaurant. She couldn't countenance not working. She loved to mix with new people and made friends easily. When she was well she loved the restaurant job. It suited her gregarious personality. But as soon as her boss realized that she had epilepsy Eleanor felt that she was being regarded with suspicion.

"My seizures weren't so bad then," she told me, "but I think they wanted to fire me as soon as they found out. They saw me fall one day and then they just stopped trusting me."

One day at work Eleanor had picked up a hot plate, stumbled and spilled some of the food near a customer. Eleanor knew what seizures felt like. She woke in the morning and could tell immediately if she was going to have a bad day. If that happened she avoided serving and asked to be moved to other jobs. On that day she knew she hadn't stumbled because of a seizure. Still the manager of the restaurant called her to a meeting and suggested that she might wish to leave. Eleanor had to fight to stay. Having epilepsy, even without seizures, threatened things for her. She had learned to live with the uncertainty of the disease but other people couldn't necessarily do the same.

A year later Eleanor chose to resign. Her seizures were becoming more regular and she knew she was unsafe. She struggled to get another job and, hating to be idle, became a play volunteer in a pediatric hospital. She loved children and could spend happy hours sitting on the ground with them.

"I can't pick them up. Or run after them. But there are always nurses and other staff to do that for me."

Having her own children was and is something Eleanor thinks about from time to time. She alternates between wanting it, being excited at the prospect of it – and then feeling terrified about how she would cope.

In the last couple of years Eleanor's seizures have become so frequent that she can't walk unsupported. There always has to be somebody there to catch her. I would like Eleanor to use a wheelchair. At least then she'd be safe. She fears that the moment she stops walking she will never walk again.

"Anyway, I'll still have seizures in the wheelchair. I won't be able to push myself. Somebody else will have to push me like I am an invalid."

With or without a wheelchair Eleanor would need somebody with her. Being able to walk, however precariously, maintains control and normality in her life.

There have been accidents. Once Eleanor fell through a glass door. Wide awake for the whole episode she saw the glass come towards her and felt it splinter around her. She landed on broken glass. There were shards in her hair, and in her clothes. She was miraculously uninjured. She got up, dusted herself off and hasn't let it stop her.

She had another near miss on holiday. Her sister was taking a photo of Eleanor posing in her bikini in the sea. Her sister is reliable and loyal. She is also human. For some reason that they both forget now her sister looked away for a moment. Eleanor was waist deep in water. She had a seizure and flopped forward landing face down.

"I knew everything that was happening but I had no control over my body. My head was underwater and I was breathing it in. I felt myself drowning."

A man standing nearby noticed her and lifted her head up. A minute later she walked out of the sea as if nothing had happened.

"I thought I was going to die."

These days Eleanor, her family and friends have to be one hundred percent vigilant at all times. Eleanor has started to wear a helmet. She is rarely alone. She doesn't eat unless somebody is watching her. She goes up and down stairs on her bottom even if she is perfectly well. Just in case.

The decision to have an intracranial EEG study to test her suitability for surgery is the sort of choice that a person makes when there is no other option.

When I called to see her on the ward after the electrodes had been put in I was shocked. She looked terrible. One side of her face seemed swollen. I could have touched the wires that led directly to her brain. The brain has no sensation. She couldn't feel the electrodes recording her brainwaves but she could feel the after-affects of the anesthetic and the aftermath of having had her skull opened.

"I've had loads of seizures already," she said.

"I know. The surgeon will take out the electrodes later today."

"How soon will you decide if I can have surgery?"

"It'll be a few more weeks, I'm afraid. We have to have another multidisciplinary meeting and discuss all the results again, taking this recording into account."

"My life is on hold. Why does it have to take so long?"

Eleanor was waiting to get better. She never stops believing she will.

A week later, we had that meeting. The extent of the tumor was uncertain – the scan did not show sharply demarcated edges and it was thought to extend beyond what was seen on the scan. The tip of the iceberg. The intracranial electrodes had shown the seizure discharge as beginning in an area treacherously close to the motor region for the leg. The tumor extended as far as the primary motor cortex in places. Only part of the tumor could be removed. Whether or not that would work was debatable. Eleanor met with the surgeon after that discussion.

"She says I can't have surgery," Eleanor told me when I saw her next. "I am able to walk now and if I have the operation it could paralyze me and then I might never walk again. And the chances of the surgery working are only twenty percent. I might have the operation, end up in a wheelchair and have just as many seizures as before."

Eleanor was upset and just a little angry. It had taken two years of scans, discussions and an invasive brain operation for us to discover that surgery was too dangerous for her.

"I'm sorry," I said.

"I know you were trying to help."

"Things are changing every year. In the future there may be safer ways to operate."

I felt I had to offer something. Some hope. Eleanor was having a hundred seizures a day.

"I'll wait because I have to – but please, I don't want to try anything new for a while. No drugs, no surgery, nothing."

"Okay."

12

MARION

The life so short, the craft so long to learn.

—Hippocrates (*c*.460–*c*.370 BC)

There was a period during my time as a junior doctor when I panicked whenever my bleeper went off calling me to the intensive care unit. I knew the call would be about Marion; for weeks every call had been to say that Marion was in crisis. I was not the only junior doctor who dreaded hearing Marion's name. I suspect the consultant felt the same. Every treatment he'd suggested so far had failed. Marion's seizures pressed on regardless. We were pouring drugs into her and her brain was reacting as if the drugs were as inert as water.

Marion was a senior staff nurse at a regional hospital. I never met her when she was well but I was told that she was a clever woman, genial and uncomplicated. Thanks to her the thirty-bed medical ward that she oversaw ran smoothly. She had a reputation for being serious-minded and even-handed. When she began to show features of depression the people who knew her well had been very surprised. It was quite uncharacteristic for her.

The problem had started two months or so before I first met her. Colleagues at work noticed that Marion had become quiet and socially withdrawn. She had cried several times on the ward in response to the challenging behavior of patients. Usually she would have been the calming voice of reason.

The problem evolved very quickly. Low moods began to alternate with periods of absolute ebullience. She talked non-stop. She developed grandiose, impossible ideas for developing the scope of work on the ward. She also started drinking heavily. Often after work she corraled others to join her in the pub opposite the hospital.

Confused colleagues didn't suggest that she seek medical help until the hallucinations started. First she reported hearing voices. People were whispering to her constantly, she said. Then she saw rodents and snakes. The voices upset her but the visual hallucinations were frightening and caused her to become agitated. Friends who knew her well pressed her to see her GP. She got angry. She simply could not recognize that the things happening to her were anything but real.

One day at work Marion could not stop talking. She fought with a friend and colleague who tried to suggest that there was something wrong. At the end of the argument she slapped her friend before becoming uncontrollably excited and disturbed. She paced the corridor. Nobody could get through to her. Ultimately Marion was led by a security officer and two friends to the casualty department. She was seen by a psychiatrist who said she was having an acute psychotic crisis. She was sectioned – committed against her will to a secure psychiatry ward.

From the outset the psychiatry team thought there was something unusual about Marion's psychosis. It was explosive in onset. Her intellect seemed to be involved, and her memory in particular. She didn't recognize people she knew well. They arranged for her to have a CAT scan of her brain and it was normal. Drug screens and blood tests were also normal. She was given antipsychotic medication. It calmed her but didn't impact on her hallucinations.

After a few days medical staff noticed that she had developed tics. He face and shoulder twitched intermittently. She grimaced and hyperventilated multiple times per hour. She became fidgety. She couldn't sit still. A neurological opinion was sought. The ward was waiting for the neurologist to arrive when Marion collapsed and had a seizure.

The psychiatric hospital was miles from the local medical hospital. An ambulance was called and Marion was rushed to the accident and emergency department. While waiting for the ambulance and during her journey, Marion had recurrent seizures. Periodically her whole body convulsed, then stopped, then started again. Drugs given by the paramedics and then by the emergency doctors didn't help. Marion was transferred to the intensive care unit. Once there an EEG confirmed the continuous brainstorm of status epilepticus. Marion was given propofol, a sedating drug that suppressed the seizures and kept her deeply anesthetized. She was put on a ventilator. Her blood pressure was controlled with drugs.

Marion ultimately stayed in the intensive care unit for six months. Every attempt to wean her off the sedative drugs resulted in seizures. Epilepsy drug after epilepsy drug was added. Eventually the seizures did reduce in frequency, although it was never clear if the drugs had made the difference or if the underlying disease had burned out on its own. Whichever the case, Marion was not left severely disabled.

Throughout her stay Marion had undergone every available investigation. Her first MRI scan was normal but subsequent scans showed swelling in both temporal lobes. A lumbar puncture was done to look for clues in the cerebrospinal fluid (CSF) that bathes the brain. It showed a mild inflammatory response. She was diagnosed with limbic encephalitis, which means inflammation of both limbic areas on the medial surface of the temporal lobes. It's more a description of the problem than an explanation.

The herpes virus can cause encephalitis and has a predilection for the temporal lobes so she was treated for that. Her blood and CSF ultimately tested negative for any herpes infection and the antiviral treatment made no difference. Marion even had a brain biopsy. The surgeon took a piece of brain from the swollen right temporal lobe. The pathologist declared it normal brain tissue.

"He must have biopsied the wrong spot," we concluded.

Throughout her stay in intensive care I was called to see Marion regularly. Sometimes she had facial twitching and nobody knew if it was a seizure or not. In response to that her sedative medication was usually increased. At other times she was suffering from complications of the treatment. A urine infection. A chest infection. An allergic reaction. Constipation. Distended abdomen.

As often as possible the anesthetists tried to wake her up. I was usually there for that. Bit by bit the sedative dose was reduced. Sometimes it was stopped and she woke up fully.

When she did she would battle against the tubes and intravenous lines. The nurses had to settle her. If she was lucky she got to spend a day or two off sedation. Inevitably it didn't last and the seizures restarted. More often she didn't even get those days off the ventilator. As soon as the dose of propofol was reduced her face began to twitch and she went straight back into a generalized convulsion.

I hated being called to see Marion. She was not that much older than me, her life was similar to mine. She made me feel vulnerable. I didn't know what to do for her and felt out of my depth.

"Do you think she's seizing?" the nurse in the intensive care unit would ask when Marion began grimacing or twitching.

I didn't know.

Marion did wake eventually. The status epilepticus stopped, although the seizures continued sporadically. She moved from intensive care to the high dependency unit. From there she made it to the medical ward and eventually to the rehabilitation unit. In all she spent almost a year in hospital. She was utterly diminished by her illness. Both temporal lobes were scarred. Her hippocampi were shrunken. Huge chunks of memory were missing. Everybody visiting Marion had to introduce themselves clearly. She couldn't remember new people unless they reinforced their identity at every meeting.

Marion's career was over. Her ability to learn new things was almost obliterated. She became exceptionally anxious and experienced rapid shifts in her mood. Tiny things upset or frustrated her. The calm woman in charge that she had been no longer existed. One of the things that hurt the most for those who knew Marion was the memory of how she had been. When I saw her in the rehabilitation unit she had pictures of her old life all over her room. Her family had put them there in the hope that they would jolt her back to some sort of normality. They never would. Her brain was damaged and that was that.

Once when I called to check on her Marion proudly showed me a picture taken at her graduation. She was wearing her nursing uniform.

"I was a nursing sister, you know," she told me, "I managed a medical ward at St Christopher's."

I knew this already, of course. But I don't think it was me that Marion was reminding. It was herself.

*

In the past twenty years I have seen many patients go through a very similar illness to Marion's. Most neurologists have. Usually the people affected are young. Often, although not always, they are women. Out of the blue they develop catastrophic seizures, psychiatric problems and temporal lobe inflammation.

For years nobody had any idea what was causing these young people to become so dramatically and irreversibly brain-damaged. Lots of people like Marion had their lives changed beyond recognition. Or lost their lives. Then in 2007 a scientific discovery finally made sense of this sort of limbic encephalitis. The anti-NMDA receptor antibody (NMDAR ab) was identified. N-methyl D-aspartate receptors are found in the brain and act as gates that control the flow of ions in and out of cells. This affects the electrical excitability of the cell. NMDA receptors protect the health of neurons and are thought to play an important role in memory. An antibody is a type of protein produced to help fight external pathogens such as viruses. NMDAR ab is an autoantibody. Instead of being directed against something external it is directed against an individual's own NMDA receptors. NMDAR ab is one of several antibodies recently identified and shown to cause life-threatening encephalitis like Marion's.

The discovery of a mechanism for limbic encephalitis has benefited many people. In some the antibody is produced as a result of an underlying tumor, particularly ovarian tumors. Find the tumor and remove it and the antibody goes away. Not everybody has an identifiable cause for the errant antibody, however. In those where there is no tumor the treatment is to suppress the immune system. Prompt diagnosis and treatment has given people like Marion a hope of recovery they would not have had before 2007.

The brain remains a mysterious organ, and brain disease is as challenging as it has always been, but there has been progress. Neuroscience is beginning to put the puzzle together one piece at a time. However, in the practice of clinical neurology, huge scientific discoveries can feel very tiny to a front-line doctor treating patients. It takes a long time for scientific discovery to become of practical help. Brain disease can be as incurable as ever and we have no ability yet to restore lost function. But everything we learn about the healthy brain offers some insight into why things go wrong.

There are big changes coming. In 2013 the Human Brain Project was started. It is a collaborative project aimed at bringing global researchers in all fields of neuroscience together in the hope that it will speed the rate of progress. One aspect of its work has been to create the Big Brain. This is a high-resolution 3D atlas of a human brain. It is the twenty-first-century version of Brodmann's areas and Penfield's homunculus. The atlas has been created from sections of the brain of a sixty-five-year-old woman who passed away from a non-neurological problem. Her brain was sliced into twenty-micrometer slices (one micrometer is one-thousandth of a millimeter) and each was stained and photographed. The Big Brain was then constructed from those slices. MRI can see to a resolution of one millimeter so the Big Brain has produced a magnified picture of the brain more detailed than ever before.

Of course it is treatments that we want most urgently. Some of the most recent discoveries could lead to those. The field of genetics is probably where the greatest hope lies. We now know that genes have on and off switches which can be affected by external factors. If they could be harnessed for medical use disease could be stopped before it had even started. Neuroplasticity is also an intense area of research interest. Since the brain has the capacity to make new connections we need only learn how to promote that ability and it could open the door to recovery even for people with severe brain injury.

There is no doubt that there be will new surgical procedures. Brain surgery could be unrecognizable in the very near future. Computerized navigational systems already allow surgeons to focus on very specific areas of the brain to ensure that surgery is as conservative as possible. But with the minimally invasive techniques currently in development they may soon be able to operate without even opening the skull. Computer-guided lasers or heat or ultrasound could be used to destroy the offending brain area while leaving the surrounding healthy brain untouched.

Even more exciting is the development of robotics, an example of science exploiting what it knows about the normal physiological mechanisms. The brain sends its messages by means of electrical signals. Mind-controlled prosthetic limbs can tap into those signals and allow the user to both move the limb and detect sensation. To see robotics in use is miraculous.

It's exciting to read about every new innovation, but my optimism is always contained. Every outpatient clinic and ward round gives me several reasons to come back down to earth. Most of the advances have been in our understanding of the fundamentals of how the brain works, how it develops, how it is organized. Practical applications of those discoveries are less forthcoming. The lack of both cure and prevention for problems like multiple sclerosis, epilepsy, Parkinson's disease, Alzheimer's, autism, schizophrenia and many more is very sobering.

Sometimes I can't shake the thought that I am very little help to a lot of my patients. It took me years to discover the brain lesions that were causing August's running attacks. Even after I had, I couldn't offer her a cure. Her seizures are only fractionally better now than when we first met. If asked, I think I would say that I have been useless to her – even though I know that August doesn't see it that way. The rapid medical advances of the last twenty years and the intense focus they place on disease and cure have made me feel I am only doing my job well if I always get the diagnosis right and if I make everybody better. But neither is possible for a doctor working with brain disease. But still Ray and Mike and Adrienne and others have all expressed their gratitude for the help I have given. Even when that help has been ineffective, even when it has made them worse, they have been grateful for the effort. It surprises me every time, because I keep forgetting that my job isn't about treating disease, it is about treating people. There is far more to practicing medicine than heroic procedures that save the day. There is far more than just cure.

The advent of modern technology and medical advances is exciting. But making a diagnosis in neurology still depends heavily on the flavor and details of a story, on the comparison to other patients, and on intuition. Exploring the brain still relies on the help of individual patients as much as it does on any type of scan. Historical discoveries were often attributable to particular named patients – Phineas Gage, Tan, Henry Molaison. That hasn't changed. The Big Brain belongs to a single woman. She lived without ever knowing how important her legacy would be. Perhaps in the future her name will be as famous to neuroscientists as those others are. For everything we don't know, there may be a patient (a person, not their scan or their blood test) waiting to give us the answer.

I learn from my patients every day. Many of the stories told by the people in *Brainstorm* were an absolute mystery to me when I heard them first. My neurology textbook had no chapter to help me to explain Eleanor's odd symptoms. The method I used to understand her problem was just the same as the clinical–anatomical method described by the nineteenth-century physicians. Listening, watching and time were my diagnostic tools. But the very least of what people like Eleanor have taught me is about epilepsy and brain anatomy. Much more than that they have shown me what resilience means, and about how to live life well despite all the challenges it proffers.

There are still gaping holes in our knowledge about the brain. Even the basic questions remain unanswered. We still don't know why we sleep or the purpose of dreams. We don't know how the brain creates intelligence. Or consciousness. Or how free will is created. We are still caught up with the rudimentary building blocks and quite far away from answers to these much bigger questions. For my part, I am not even certain that I want every single question answered. If we knew everything about how the brain functioned, what would we be then? Just sophisticated computers? Machines that could be reprogrammed? For all the people who allowed me to tell their stories in this book I want a cure. For the future I want to know how to prevent diseases. These would be advances enough for me. I have no need to see humanity unraveled; it is enough to observe it. But, of course, I needn't worry. We are not even close.

ACKNOWLEDGMENTS

First and foremost, huge thanks must go to all the incredible people who allowed me to tell their stories in this book. Your generosity and strength is an example to us all. For the purpose of concealing people's identities I have changed some parts of some stories, but never the essential medical details or the heart of the person.

Thank you to the neurology teams I have worked with – especially Adele Larkin, Jennifer Nightingale and neurophysiologist extraordinaire Fiona Farrell.

As always I am indebted to Becky Hardie of Chatto & Windus for her endless patience and wisdom. Thank you to David Milner whose brilliant eye for detail is astonishing to observe. Also to Matt Broughton for his wonderful cover design that I loved at first sight. ("But why is there a rabbit on it?" my sister asked. You have to read the book to find that out!)

My agent Kirsty McLachlan opened the door onto the writing world for me. That is something I will never forget and for which I am always appreciative. Thank you for your ongoing support.

Every terrified, self-doubting writer needs a writing support group. Mine comprises Gemma Elwin Harris and Jenny

Johnson. Here's to the cosy pub chats, delicious dinners and critical book talk. I look forward to reading the fruits of your labors.

And, finally, apologies to the family and friends who have tolerated my erratic availability and constant complaints about "deadlines."

INDEX

SUZANNE O'SULLIVAN, MD, is the author of *Is It All in Your Head?* (Other Press), which won the 2016 Wellcome Book Prize. She has been a consultant in neurology since 2004, working first at the Royal London Hospital and currently as a consultant in clinical neurophysiology and neurology at the National Hospital for Neurology and Neurosurgery, as well as for a specialist unit based at the Epilepsy Society. She has developed expertise in working with patients with psychogenic disorders, alongside her work with those suffering from physical diseases such as epilepsy.

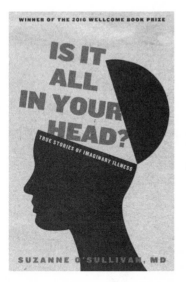

Also by Suzanne O'Sullivan, MD

IS IT ALL IN YOUR HEAD?
True Stories of Imaginary Illness

A neurologist's insightful and compassionate look into the misunderstood world of psychosomatic disorders, told through individual case histories

"Doctors' tales of their patients' weirder afflictions have been popular since Oliver Sacks...Few of them, however, are as bizarre or unsettling as those described in this extraordinary and extraordinarily compassionate book."
—JAMES McCONNACHIE, *Sunday Times*

WINNER OF THE 2016 WELLCOME BOOK PRIZE

It's happened to all of us: our cheeks flush red when we say the wrong thing, or our hearts skip a beat when a certain someone walks by. But few of us realize how much more dramatic and extreme our bodies' reactions to emotions can be. Many people who see their doctor have medically unexplained symptoms, and in the vast majority of these cases, a psychosomatic cause is suspected.

In *Is It All in Your Head?* neurologist Suzanne O'Sullivan, MD, takes us on a journey through the world of psychosomatic illness, where we meet patients such as Rachel, a promising young dancer now housebound by chronic fatigue syndrome, and Mary, whose memory loss may be her mind's way of protecting her from remembering her husband's abuse. O'Sullivan reveals the hidden stresses behind their mysterious symptoms, teaching us that "it's all in your head" doesn't mean that something isn't real, as the body is often the stand-in for the mind when the latter doesn't possess the tools to put words to its sorrow.

OTHER PRESS   *www.otherpress.com*